Praise for
The Permaculture Transition Manual

Finally a resource that holistically connects the dots between all the elements of permaculture. *The Permaculture Transition Manual* reaches beyond the design and planting side and serves as an inspiring guide to craft a lifestyle rooted in permaculture principles, from cooking in the kitchen to cultivating community.

—Lisa Kivirist, author, *Soil Sisters: A Toolkit for Women Farmers*

Drawing on his experience as both a teacher and permaculture practitioner, Mars presents a complete blueprint for anybody to make their rural property or suburban home more beautiful, economical, and resilient. He also offers inspiring ideas to de-clutter and make the most out of smaller urban spaces.

—Erik Curren, Publisher, *Transition Voice*

Definitely a book for these crucial times, *The Permaculture Transition Manual* discusses all the steps necessary to initiate positive change. Ross Mars covers all the bases – both new converts and readers already conversant with permaculture will find suggestions to complement and facilitate their journey towards a more sustainable existence. The writer's scientific background does not prevent him from keeping the information accessible to all with point by point analysis of what's required to halt the devastating ecological freefall that's been instigated by the profligate lifestyle of today's society.

—Jenni Blackmore, author, *Permaculture for the Rest of Us*

Holistic solutions rather than piecemeal fixes are required to address the many converging crises in our world. *The Permaculture Transition Manual* reviews the problems we face, then builds on the view from 30,000 feet with an excellent exploration of the permaculture design process, peppered with information on compost, graywater, alternative energy, and much more. Ross Mars has written an excellent introduction for the beginner!

—John Wages, Publisher, *Permaculture Design* magazine

The Transition and the Permaculture movements offer road maps to a future that will not only be sustainable, but will allow people to thrive. Ross Mars is the perfect tour guide with *The Permaculture Transition Manual*. Offering a comprehensive and integrative approach to living that anyone can incorporate into their lives, this book offers a treasure-trove of creative practices and vital information. This is the book I would want on a desert island as a survival guide. It is a great resource for any community or individual looking to create a better future.

—Hannah Apricot Eckberg, editor,
Permaculture Magazine, North America

the PERMACULTURE transition manual

A Comprehensive Guide to Resilient Living

ROSS MARS

Foreword by ROB HOPKINS
author of THE TRANSITION HANDBOOK

Illustrations by Simone Willis

Cover design by Diane McIntosh. Cover Images © iStock

Illustrations by Simone Willis, additional illustrations: Chapter 9 — Kristin Scali.
Celery © adobestock_104819563.

First published as *How to Permaculture Your Life.* Copyright © Ross Mars
First published in Australia in 2015 by Candelight Trust, 4110 Phillips Road, Mundaring, WA 6073
www.cfpermaculture.com.au

Published in North America by New Society Publishers under license from Permanent Publications,
The Sustainability Centre, East Meon, Hamsphire GU32 1HR, UK www.permaculture.co.uk

Printed in Canada. First printing October 2016.

Paperback ISBN: 978-0-86571-835-7 Ebook ISBN: 978-1-55092-630-9

Inquiries regarding requests to reprint all or part of *The Permaculture Transition Manual* should be
addressed to New Society Publishers at the address below. To order directly from the publishers, please
call toll-free (North America) 1-800-567-6772, or order online at www.newsociety.com

Any other inquiries can be directed by mail to:
New Society Publishers
P.O. Box 189, Gabriola Island, BC V0R 1X0, Canada
(250) 247-9737

Library and Archives Canada Cataloguing in Publication

Mars, Ross, author
The permaculture transition manual : a comprehensive guide to resilient living / Ross Mars ;
foreword by Rob Hopkins, author of The transition handbook.

Includes index.
Issued in print and electronic formats.
ISBN 978-0-86571-835-7 (paperback).--ISBN 978-1-55092-630-9 (ebook)

1. Permaculture. 2. Organic gardening. 3. Agricultural
ecology. I. Hopkins, Rob, 1968-, writer of foreword II. Title.

S494.5.P47M37 2016 631.5'8 C2016-904056-9
C2016-904057-7

Funded by the Government of Canada | Financé par le gouvernement du Canada | Canada

New Society Publishers' mission is to publish books that contribute in fundamental ways
to building an ecologically sustainable and just society, and to do so with the least possible impact
on the environment, in a manner that models this vision. www.newsociety.com

CONTENTS

FOREWORD

PERMACULTURE IS A SYSTEM OF DESIGN for people living within nature. It has become common practice in many countries since it was developed forty years ago. However, we live in a world that is rapidly changing and we need to respond positively to this. Permaculture is now central to bringing about that widespread change, even more so with the advent of the Transition movement.

Transition started as a conversation. From that initial conversation between a small group of people, who had wide and diverse backgrounds, came a phenomenon that is now found throughout the world. We are seeing communities around the world adopting and adapting the tools that can enable our communities to reengage with each other and to build resilience.

This book, *The Permaculture Transition Manual: A Comprehensive Guide to Resilient Living*, written by well-known teacher and author Ross Mars, is aptly named. It is full of practical ideas, techniques and strategies that are built on sound scientific and ecological principles.

Over the last forty years Ross has researched, observed, experimented and developed integrated systems, and he shares his knowledge here. His work took a different turn when he became interested in permaculture 25 years ago. This book has been a long time coming. Here is a toolkit of useful plants; alternative farming systems that repair and build soil; skills to add value to your produce; strategies for energy and water efficiency; and techniques to enable us to re-skill and relearn arts and crafts many of us have lost in recent generations. Ross has shown that we do have the knowledge and technology to allow us to live in a lower carbon future, that we can grow nutritious fruit and vegetables, even in small spaces, and we can have a lot of fun and enjoyment doing so. This is not so much a transition *away* from something, as a transition *towards* something eminently more nourishing and healthy.

These innovative ideas will enable us to farm profitably even in extreme climates and landscapes, undertake tasks and build structures we thought were difficult to do, and select plants for specific functions and regions. This book is full of immensely useful information and handy hints for those of us who share a vision for a better future. Using the Earth's resources more wisely and gently is the only way we might be able to leave our world in a better condition for our children and grandchildren.

Although Ross is recognized worldwide as a wastewater specialist, his focus has always been to see how natural cycles and the interrelationships in ecosystems can be improved and utilized to develop more sustainable ways of living on this planet. This book continues and deepens that work.

In her book *This Changes Everything,* Naomi Klein wrote "there are no non-radical solutions left" if the world is serious about avoiding catastrophic climate change. This book sets out what some of those solutions will look like, and the many benefits they will bring us. It took huge creativity, hard work and imagination to create the Industrial Revolution and we have much to thank its designers for. Yet the path forward from here will look very different, but there is no reason that the urgent dash away from fossil fuels won't unleash the same levels of creativity and ingenuity. Permaculture design principles deserve to be at the heart of that, and this book offers us a powerful crash course in the thinking we will need if we are to create a future that actually works for everyone.

Rob Hopkins

Founder of the Transition movement
and author of *The Power of Just Doing Stuff*

PREFACE

THIS IS NOT A BOOK solely about techniques to enable people to become better gardeners, nor is it a textbook for a Permaculture Design Course. There are already many books that fill those roles. This is a book about strategies, some techniques and useful information about plants, technology and materials. As a scientist, I am interested in facts, numbers and statistics, so you may be surprised to read about only particular plants, and I have endeavored not to make too many unsubstantiated claims. I have tried to present factual information rather than fable, provided some elaboration on how things work in the world, and attempted to make a complicated and complex world a little easier to understand.

DEDICATION

THIS BOOK IS DEDICATED to two Bills and two Davids. The first Bill is my late father Bill Mars. Dad used to make me weed the vegetable patch when I was young, which, strange as this may seem, instilled in me a love for plants and getting my hands dirty. We always had vegetables, a few fruit trees and fresh eggs from "the girls." This is the path I have followed throughout my life.

The other Bill is, of course, Bill Mollison, who continues to inspire me and several generations of people throughout the world to become better conservationists.

The first David is my grandfather David Williams. When we used to visit as a young boy, I would rush out to the backyard and either head for the ramshackle tool shed to potter around, build things (but mainly destroy things) or forage in the garden, which was full of fruit trees and vegetables. I still have some of Grandad's hand tools.

We are all also indebted to fellow West Australian (and Fremantle-born) permaculture author and teacher David Holmgren, who traveled to Tasmania so long ago now and became the co-originator of the permaculture concept.

SETTING THE SCENE 1

What is changing? Is there a problem?

MUCH HAS BEEN ALREADY WRITTEN about our changing world and this book does not delve into the causes, the facts or the solutions for climate change, peak oil and food sovereignty.

These are the "big three" that are taking center stage at this point in history, and for those of you who are not aware of some aspects, here is a short summary of each.

People often focus on one problem, so you find groups solely focusing on peak oil or climate change, or some other ecological problem. We really need to broaden our thinking in a more holistic way, because when you consider human existence on this planet, I believe that the most pressing issues are food sovereignty (some suggest food security), peak oil and then climate change, in that order, although this order is debatable.

While change is all around us, it is our response that is most important. As we all move through this period of transition, we will need to re-skill and adapt, find new ways to solve new problems, and amongst all of this, enjoy our lives and our families.

Climate change

When we first learned about our Earth heating up a few decades ago the focus was on "global warming." There was contention, as there still is today, about whether global warming is caused by humans, enhanced by humans or just a part of nature's cycles — as historical evidence has shown cycles of ice ages and warmer climates in the past.

As science and observational data developed in recent years, it became more apparent that the focus was about "climate change," the predictions of severe weather events and extremes, changing rainfall patterns and changing

Severe weather events will become more commonplace.

Fossil fuels are being rapidly consumed, and the gases produced from burning contribute to global warming.

temperatures. While a few skeptics still believe humans haven't really had any effect on our climate, it is commonly held that we have increased the rate at which Earth changes are occurring.

From a permaculture perspective, this is a real concern and we need to respond to this in a positive way. We need to grow food plants that are resilient to local change and to develop techniques for an integrated pest management strategy, as the cycles of pests and predators also change with the seasons and some of these are "out of kilter" with each other.

Coupled with our changing weather patterns are increasing levels of pollutants in our atmosphere. We are still pumping tons of carbon gases into the air every day, and this is overlooked by governments and industry, who both seem complacent about the destruction we are doing to the planet.

It would be true to say that some governments exhibit deliberate inaction, and this would only make the impact of climate change worse.

» DID YOU KNOW?

If you want to know more about what this means for our future, make a search online for transition towns, peak oil, 100 mile diet, post carbon cities and the relocalization network.

Peak oil

Oil resources are finite — we will eventually run out. We have so far used about half of the known reserves in the world (some scientists suggest more than half), but as more oil is pumped from the ground, what is left gets progressively harder and more costly to access. Fuel prices rise and the cost of synthetic chemicals (both fertilizers and pesticides) increases dramatically as the price of oil increases. This has implications for the cost of food.

Like climate change, large numbers of books and articles have been written about the effect of peak oil on human existence. Our modern society is totally dependent on oil, that relatively cheap fossil fuel that drives the economy, fuels machinery and transport systems, and is used to manufacture a large number of products.

While we still have enough oil and gas and coal to last decades, we need to develop energy alternatives.

Many writers refer to this period of energy decline as a time when we could wean ourselves off oil and develop longer-lasting energy sources, such as wind, solar and water.

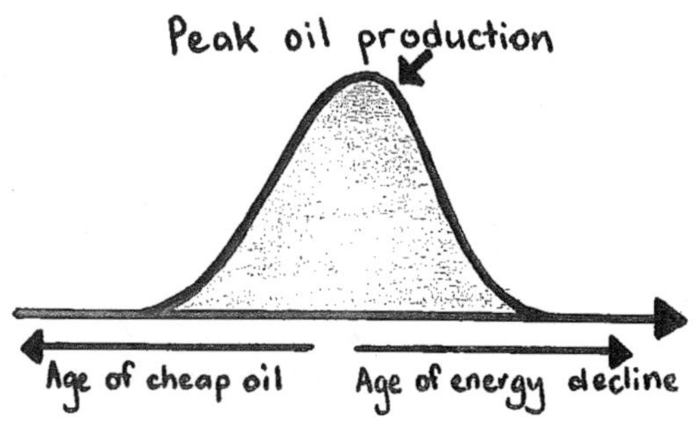

Peak oil has occurred. We are heading towards the age of energy decline.

Bicycle use throughout the world has increased in recent years.

» **DID YOU KNOW?**

The concept of peak oil was put forward by Marion King Hubbert in the mid-1950s. He predicted that we would rapidly use the world's vast oil reserves and that the rate of extraction would "peak" when half of the reserves were consumed. We appear to be at the halfway mark, and now we are heading down the slope.

As oil is depleted, goods and products will increase in cost. You might be surprised to find out just how many things we have in our homes that use oil in their manufacture. There will be issues around transport and food production, the cost of goods and services, and production of electricity and heating in our future.

I would like to think that the impact of peak oil can be offset to some extent by new and improved technologies and by changes in the behavior of individuals.

Food sovereignty

First we need to distinguish between food sovereignty and food security. Food sovereignty is about our control over the food we eat. This means that as we are the growers and consumers of food, we should take more responsibility and be more involved in the policies of food production and its distribution, rather than the large corporations that currently dominate the global food system.

Food sovereignty is about the rights of people to make their own decisions regarding the way they farm and what they grow, and this enables greater choices for consumers.

Food security is about the supply of food: year-round access to sufficient food that is safe and nutritious and can support an active life. It is both about quality and quantity. Food security has four components: availability, access, use and stability.

Food availability relates to production and distribution: how much food can be grown and how it can be moved to other areas. A country doesn't necessarily have to grow food in order to achieve food security, but most countries endeavor to promote agricultural enterprises.

Access to food refers to the ability (or inability) to obtain food. Poverty can limit access to food, so income is strongly correlated with the purchase and allocation of food.

Food use refers to the efficiencies of individuals in metabolizing food, so both the food quantity and quality are important. People need to eat safe foods and to make sure their dietary requirements are met.

The fourth component of food security, stability, is about our vulnerability in regularly obtaining adequate food. It is a reflection of the previous three components (availability, access and use) over time.

As we are losing more arable land each day, and as the world's population keeps growing, eventually there won't be enough food for everyone. This becomes even more severe as our climate changes (and so plant cycles change, and drought or fire may seriously affect some food bowl areas) and the cost of operating machinery (costs of production and fuel) escalates.

No amount of science, genetic engineering, cloning or advances in agricultural techniques and methods will provide the shortfall.

Many people believe that to grow more food on the scale required will ultimately mean the increased use of pesticides and herbicides to control the pests that are just as keen to eat our food as we are.

Furthermore, half of every ton of fertilizer applied to fields never makes it into the plant tissue. It ends up evaporating or being washed into local watercourses or leached through the soil.

» DID YOU KNOW?

In Australia 440 different pesticides are routinely used, and in 40 years of intense pesticide use we have not eradicated a single pest.

This madness must stop. If we are to have a future, we need to grow our food organically. We need to embrace the concept of food sovereignty and work towards more recognition and support for our farmers and growers. But we also need to address the issue of overpopulation. There just won't be enough resources for everyone, and while governments and countries recognize this problem, few seem to be doing anything about it. Some countries have brought in measures to curb population growth and they have been looking afar to ensure they secure resources to meet their growing needs. Governments move forward in an exceptionally slow way. Any changes that occur will happen at the grassroots level, and this is why the Transition initiative and other similar endeavors have been so successful to date.

Food sovereignty is about controlling where your food comes from and what types of foods you want to consume.

The issues surrounding food sovereignty and food security are complex because they involve international trade and economic development, and impact directly on our health and wellbeing.

Some people suggest that there is enough food in the world to feed everyone adequately; the problem is distribution. Others contend that we need to develop strategies for resilience because we just don't know

We need to learn how to better use foods.

Permaculture promotes home food production.

what our food production capacity will be in years to come. We need to learn how to make better choices when buying food and make better use of leftovers and thus reduce food waste.

Historically, permaculture focused on redesigning our agricultural systems and reintroducing food production into the household economy. This led to excess produce being stored and, in turn, greater food security.

We believed that food security was storing enough away for tougher and lean times, when particular seasonal foods were not available. And while we still hold onto those beliefs, we also need to see the bigger picture.

The challenge we need to set ourselves is:

- How can we strengthen and expand our current farming sector, as many industrialized countries are experiencing declines in their annual food production?
- How can we support the agricultural sector of our community to produce safer foods?
- How can we make farming a more resilient enterprise because, as we have seen so often, it only takes a severe drought or extensive floods to seriously cripple a country's food production?

These questions are clearly about food sovereignty as they involve economics, markets, policy, politics and people.

The power of community revolves around food. Food is what brings people together.

People get and understand food much more easily than water and energy efficiency, so it is easier to bring people together to discuss growing food. That is where permaculture comes in.

SUSTAINABLE LIVING PRACTICES 2

Sustainability

SUSTAINABILITY HAS BEEN DEFINED as those activities and resources that can be maintained at a level that won't compromise current and future generations from meeting their needs. This means that renewable resources such as timber, food crops and fish should not be consumed faster than they can be replaced, nonrenewable resources such as oil and gas must not be exploited until they can be replaced by renewable energy systems such as solar and wind, and waste must not accumulate — waste has to be processed, assimilated or reused.

Putting all of this simply, the most common understanding about sustainability is that it is the responsible use of resources, recycling and growing some of our own food.

I think it is reasonable to say many people are not living sustainably. If you think about the way we farm and denude the land, the problems with our river systems, the clearing of bushland for housing estates, mining, logging and increasing levels of pollution, the loss of biodiversity and the current rate of extinction of species, then maybe it's time to reflect on how each of us can work towards making a better world for all.

Once you have reached the conclusion that we are living in an ecologically unsustainable fashion then the solution is obvious: we have to adopt sustainable living. This means that we will be accountable for our actions, that we need to find the right balance between what is available and what we actually need to survive. It will involve being smarter about how we interact with the natural world, and it will mean adopting strategies to minimize our impact on the environment. We will have to work towards both sustainable consumption and sustainable production. "Sustainable living," then, is all about ecological and environmental responsibility.

Trying to live sustainably involves the intertwining of caring for the environment with social and economic considerations. This may mean that before you buy anything you might consider the manufacturer's policies and operations so that the food, clothes and items you buy are produced fairly and ethically, that their workers are not exploited, and that the resources they use are able to be replaced.

In the social context, it is also about our wellbeing, about having the right to be healthy, about making the right choices, about building resilient communities that have the ability to adapt to change.

» DID YOU KNOW?

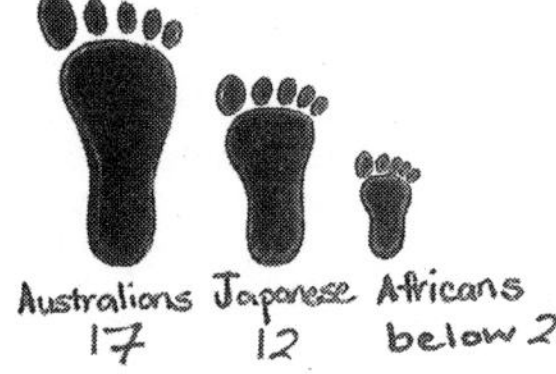

If we adopt the ideals of sustainable living, we can reduce our ecological footprint — the average area of land we require as individuals to support our current lifestyle and existence. Sadly, Australians have a footprint of about 17 acres, the USA has one of 25 acres and the UK, 12 acres. In comparison, developing countries such as India and China have footprints of about 5 acres, and Bangladesh, 1 acre.

Moving towards a sustainable future will require social change and, more importantly, behavioral change. Unless we change our worldview of what is valuable, what is really important for all life on Earth to continue, and determine what we need to change, as individuals, about our lifestyle, then no amount of talking or writing about all of these things will make much difference.

University of New South Wales academic Dr Ted Trainer argues that sustainability is not possible unless there is a zero growth economy. As we enter times of severe scarcity, he sees society moving towards small, highly self-sufficient local economies run by participatory procedures.

This is, of course, completely the opposite of what currently is in place — large corporations endeavoring to make profits for themselves or their shareholders, globalized markets and political parties and countries pushing for continual economic growth.

In this context our ideas about wealth may need to change. Wealth should not be measured by money but by good health, good food, safety and shelter, family and friends, enjoyable work, personal growth, being creative, and having time to pursue all of these things.

It was Winston Churchill who said "History tells us that we will choose the right path — once we have explored all the wrong ones. It's not enough that we do our best; sometimes we have to do what's required."

So much of the "green" work people are doing at present is really a Band-Aid solution — their thinking that setting up a community garden will change consumer society is insufficient. We need to replace the structure of society altogether.

There is a groundswell of people throughout the world who have decided to change. More people are taking public transport, riding bicycles or walking to work; using cloth shopping bags; carpooling; growing their own food — organically and without the use of pesticides and artificial fertilizers; installing solar power systems on their roof; harvesting rainwater to offset their water budget in the home; recycling graywater onto their gardens; buying green products; composting food scraps and plant material; installing water-efficient appliances and fixtures in their homes; recycling their used glass and metals; and building passive solar homes that are energy efficient.

All of these types of changes, all highly commendable, should be seen as stepping stones for the restructure of society.

However, most societies are dynamic and continually changing, and so too is our world. Sustainability implies a stable state (homeostasis) with a balance between what is used and what is replaced.

Maybe our world is changing too fast and too soon for true sustainability to be achieved. What we need is to be able to respond to change and adapt to new situations, and this is the concept of resilience, and this is where permaculture steps in.

I believe that everyone can make a difference, that we can take that first step on a long journey into the future, starting by simply making small changes in our behavior until what we do and think is second nature.

Permaculture: designing for a sustainable future

Permaculture had its origins in the mid-1970s when university tutor and mentor, Bill Mollison, and student David Holmgren developed a range of strategies with the ultimate goal of a *perma*nent agri*culture*, or permaculture.

While the underpinning foundation of permaculture was the production of permanent food crops, it developed into an all-encompassing framework about all aspects of human settlement. It became *perma*nent *culture*.

» DID YOU KNOW?

"Permaculture" was coined after J Russell Smith's book *Tree Crops — A Permanent Agriculture,* first published in 1929.

There are many books about permaculture, including a couple of my own, that deal with the basic concepts and principles on which permaculture is based. Throughout this text there will be some elaboration on permaculture design strategies and principles, but the focus here is all about practical solutions, and the relevance of permaculture in our society.

Permaculture is certainly about growing enough food and having a lifestyle that will enable you to become self-reliant (and not self-sufficient as some would believe) and less dependent on the marketplace and agencies outside of our control. But permaculture is more than this: it is about how we live, the types of houses we build, ways in which we can live more sustainably, and how we deal with water, energy, soil and living things. Permaculture is fundamentally a vocation, a way of life. It is about taking responsibility for your life and doing the things you feel are important for your own wellbeing and for the wellbeing of others and to help the environment.

Permaculture is not just about forest gardening or about mimicking natural ecosystems, although at times most of us use these phrases to explain what permaculture is in a simple sentence or two. We forget that permaculture advocates rainwater harvest, graywater reuse, renewable energy systems, and much more, all of which have nothing to do with forest gardens or natural ecosystems. We should always start our discussion about permaculture as a design approach to create edible, functioning, integrated landscapes that support all life, including humans.

Our aim should be to build productive landscapes that will take care of us while we take care of it. Permaculture doesn't have its own set of techniques, but rather a bag of tricks, tricks that are really principles and strategies to enhance vulnerable ecologies.

As we discussed in the above, concerns like global warming and climate change, food security and peak oil will have major impacts on our future survival. Permaculture is seen by many people as providing strategies to enable us to adapt to a challenging future.

How important it will be to us only time will tell, but there is a huge re-interest in permaculture throughout the world as people begin to understand how our environment is changing and how we are totally dependent on oil.

Stacking plants to increase production is one permaculture strategy.

Permaculture is a sensitive process for designing sustainable systems for living. It is based on a collection of the new and old experiences of huge numbers of people, all over the world. But it is more than a collection — it is about how each item in this collection, be it a technique, a species or a piece of knowledge, can be placed in relation to another to make the system more resilient and better able to meet the needs of the people within it.

Permaculture is often described as strategies to work in harmony with nature (which is true), but it is more than this. It complements nature and it works alongside nature to develop integrated, self-sustaining, resilient systems that can produce all of our needs.

At times we may be able to tweak nature to allow us to improve plant and animal species (mainly for our own needs, of course) and to influence our environment by large-scale reforestation projects, which, in turn, would affect water and energy movement in the landscape.

We would hope that these practices would not only benefit us but our world, while still allowing nature to provide those environmental services to benefit everything.

While permaculture books are not gardening books, gardening teaches us life skills that we can't obtain elsewhere. Gardeners tending their plants have to learn how to respond to the changes in seasons, drought, heat and cold, soil changes and bugs to ensure their plants survive and thrive. We are not oblivious to nature and we can understand what is happening around us, and respond to environmental issues as they occur.

We can be prepared for change as well as being the agents for change. The garden teaches us balance, and permaculture is also about balance — how we integrate our lives with nature, how we juggle input and output to get optimum yields, and how we tend to the needs of all the organisms present in our world so that everything flourishes.

One of the main criticisms of permaculture is the apparent proliferation of introduced plants and animals in gardens, which occasionally escape into bushland. The onus is on people practicing permaculture to carefully avoid over-simplifying the solution to our environmental and food production problems.

The belief in permaculture that natural systems and people systems have a capacity to recover from environmental degradation through the introduction of foreign species is apt, but not at the expense of losing genetic diversity or having to deal with a host of other concerns by ecologists.

An appropriate management response would be to firstly define the remnant values of an area, inclusive of endemic species, and the risk level of dispersal of an introduced species.

Bringing these issues to light is not intended to deny permaculture its capacity for adopting a wider view and therefore greater community acceptance. Ecologists, in general, would appreciate permaculture taking a self-critical approach when assessing the risk of introducing any organism into a foreign environment.

From an ethical perspective we need to remind ourselves of the importance of having an ecocentric view. This means carefully considering the intrinsic worth of all living things and their right to self-perpetuation, free from human interference. Hopefully, being ecocentric in our approach means greater cultural acceptance of permaculture. An ecocentric worldview is something that legal regulation and human-based environmental education programs have failed to achieve.

And this is where permaculture steps in again: it's about design. Permaculture practitioners endeavor to design functioning ecosystems with positive enhancements to all the organisms present.

Permaculture designs do take time to establish, but once they are implemented they become more and more productive. A larger range of useful products becomes available, the level of maintenance decreases and the system becomes more complex.

Permaculture, and the ecological framework it embraces, will give people hope and enable them to develop skills that allow us to rise to the challenges of a changing world.

Permaculture design fundamentals

There are many versions of the basic principles on which permaculture is based. David Holmgren's 12 principles have been almost universally adopted, and have more rigor than Bill Mollison's principles, which are discussed in earlier permaculture works.

Discussion on these overarching principles that guide people along their permaculture journey can be found elsewhere. I want to focus on the design process itself and what steps are taken to distil and condense the ideas and concepts that make permaculture designs so unique.

When thinking about how designs are undertaken, there are six fundamentals.

1. Observe and analyze

The first step is to observe, collect and collate data. This may include soil and water analysis, local climate, water movement in the landscape, sun angles for each season, and the wants and wishes of the human inhabitants.

From this information we sort, group and analyze trends, make predictions and determine priorities.

2. Consider needs and functions

All of the things that are placed in a design are called elements. We need to think about the needs, requirements, functions and products for each of the elements in the system.

This means identifying which special requirements the various plants and animals have, how their waste can be used as a resource, what functions we want the animals to perform, how each element can complement many others, and how human needs can be met in the design we are proposing.

3. Use patterns and make connections

The third step of the design process asks: how can we integrate all of this into a holistic system, involving communities of plants, animals and people? This is where we start to use concepts such as edge, stacking, guilds and zones.

We endeavor to determine ecological interactions between all of the living components and examine land use patterns to develop strategies.

For example, we may consider the types and shapes of garden beds for maximum production, how we can place plants to grow as much food as possible without jeopardizing their needs for sunlight, nutrients and water, and where we place everything in areas of different intensity.

4. Manage energy and use local materials and resources

Permaculture design is essentially about energy. We consider how winds and storms can be deflected and their effects minimized by sector planning, we think of ways to harvest and store energy and water, we determine what recyclable materials and renewable energy systems we can access and use (with consideration of the embodied energy of these materials), and then we consider what local resources are available.

5. Increase biodiversity and productivity

Our ultimate aim should be to nurture soil to increase both food and non-food production and to increase the biodiversity and biomass of the cultivated ecosystems we develop. This will involve a whole host of techniques and strategies, including integrated pest management; growing, harvesting and storing food and materials; and replenishing spent nutrients as plants are removed from the system.

6. Design for catastrophe

Natural disasters have always been with us, so we should prepare, within reason, for flooding, drought, fire and even earthquakes. While we don't really know how our future will unfold, we can make some very good guesses of what might happen.

Simple observations of what is happening now will reinforce our resolve to address pest and plague, energy decline, severe weather and extremes, nutrient depletion, declining freshwater sources, pollution and loss of arable land.

Our designs should include contingencies to counteract possible future scenarios and to enable resilience in the systems we construct.

Sometimes you need to be the change you want to see happen.

You know what they say about the longest journey starting with the first step, so don't underestimate the power of one. I remember Bill Mollison once said (maybe not verbatim), "I can't change the world all on my own. We'll need at least three of us."

PLANNING A GARDEN

3

OUR AIM SHOULD BE TO DESIGN AND BUILD an edible, productive landscape that is developed with long-term sustainability in mind. So what would be the indicators for a sustainable landscape? The three obvious indicators are soil health, water quality and types of plants, and we need to keep these in mind when we start the design process.

Ecological principles

Whether you are building a herb and vegetable garden or developing the whole property, the process you want to implement should be based on sound ecological principles.

The components of an ecological garden can be remembered by using the acronym PAMS WASH.

P — Plants

Plants are producers. They make food, products (oil, fiber, clothing, rubber, and an endless list of things we use every day) and oxygen (which every living thing requires).

Our observations and knowledge about plant species will determine the types, varieties and numbers of particular plants we want to use. The types of weeds present on a site can also provide helpful information.

Weeds can be used as soil indicators (to some extent) and they are nature's way of repairing damaged land. These pioneers restore the fertility of disturbed soil, so weeds should be seen as opportunistic plants rather than invasive plants.

We can keep weeding a garden every year (or after every heavy rain event) or we can plant our own pioneers to protect and build soil.

Pioneers are those plants that colonize and grow first in an area, and their role is to protect, change and build soil. Many fast-growing, nitrogen-fixing plants are pioneers.

We should implement strategies that will bring about both long-term improvement and yield to a landscape. It is important that we strive to restore damaged landscapes and make them productive — creating useful yields of food, materials and products.

A — Animals

Animals are consumers. While they eat plants and each other, they are important in every nutrient cycle. They provide the manures to feed soil life and make plants grow, they aerate and mix the soil, they provide the carbon dioxide that plants require for photosynthesis, and many are keystone species that determine the fate of an ecosystem.

» DID YOU KNOW?

A keystone species is one that plays a crucial role in the survival of the rest of the ecosystem. For example, sea otters protect large kelp fields from being eaten by sea urchins. Unchecked numbers of sea urchins would devour the kelp, so otters keep the urchin numbers low.

In forest ecosystems, large food plants (fruit and nut trees) may be keystone species as they provide food for large numbers of animals at particular times of the year when some food reserves are scarce.

M — Microorganisms

Microscopic plant, animal, fungi or other single-celled organisms live everywhere. The soil is the most biologically diverse habitat on Earth, with about 50 billion microbes in every tablespoonful.

Living soil is teeming with life.

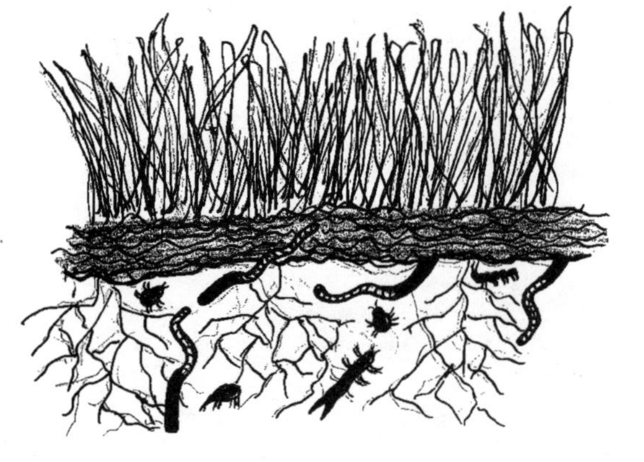

S — Soil

Healthy soil is teeming with life. If you weighed all of the earthworms, invertebrates, nematodes, bacteria, fungi and the literally millions of other organisms you would find about 2 lb in every square foot of soil.

If we assume that most of the soil life is confined to the top 8 inch then those 2 lb are found in every 60 lb of soil.

Without soil, there is no life on Earth, and this is why it is crucial that we really tend to our soils and look after them. The next chapter is devoted to creating healthy soil.

W — Water

When planning a garden one of the first things to think about is: how am I going to provide enough water to keep the plants and animals alive, and then allow them to thrive? We might need to undertake a water audit to ensure that lack of water will not limit our aspirations of a productive garden.

Sometimes we can't just rely on rain, and many fruit and nut trees may need twice the annual rainfall to produce. How we can harvest, recycle and better use water is discussed in chapter 10.

A — Air

Humans tend to have a disregard for keeping our air, water and soil clean. Clean air, clean water and clean soil are fundamental rights for everyone (and everything else too), but we seem to keep polluting these anyway.

Allowing wind to pass through our gardens can prevent many diseases, such as mildew and other molds, by maintaining good air circulation.

S — Structures

Every garden needs structures, and these typically include gazebos (patios, entertainment areas), compost bins, trellises, retaining walls, seats and garden ornaments such as sculptures and bird baths. Structures are used to support plants, protect plants, provide a habitat for animals and allow humans to sit, rest and enjoy the garden.

H — Humans

These creatures are an integral part of every ecosystem, and have the ability to change environments for the better or for the worse. We have a moral and ethical responsibility to care for the planet, and creating functional garden environments is one small step each of us can take.

So, if humans are to work in the garden they need comfort (paths), shaded walkways and a tool shed nearby, even a small potting area. Working in the garden needs to be a pleasure, so they look forward to going out in the garden for both leisure and enjoyment.

We need to keep all the humans that live there in the best health and well-being, so we need to plan for nutritional and medicinal plants, and have areas of tranquillity with appropriate seating for contemplation or meditation.

How ecosystems work

An ecosystem is a collection of interacting organisms and their surroundings. A forest ecosystem, for example, would include all the plants and animals that live there and their interaction with their physical surroundings.

Ecosystems are dynamic. They change as species move in and out, the types of plants change over time, and the soil also changes. All of this results in continual disturbance, which drives the nutrient cycles and interactions within the ecosystem. This continual change is called succession.

Ecology succession takes disturbance, just like most things in life. We need to provide or undertake the right amount of disturbance and disruption to enhance biomass productivity. Disturbance even on a vulnerable and fragile landscape can ultimately result in successful healing of the land. This may include installing dams to harvest rainfall and thinning trees to increase light penetration to a forest floor and to the soil. While nature uses disturbance in successional changes in ecosystems, we can use disturbance in appropriate ways and at appropriate times to be innovative. On the other hand, too much disturbance can result in weeds and soil damage so it is an art to get that right balance.

Provided there is enough rain or irrigation, succession drives a landscape towards forest. Landscapes tend to be patchy as disturbances (fire, storms, even clearing) always occur and interrupt the successional process. New bare ground becomes covered by weeds.

The whole significance of this is that when we garden — plant our vegetables, flowers or other (useful) plants — we are essentially planting pioneers and creating a patchwork of mini-ecosystems all at different stages.

Furthermore, as we change our soils from clays and sands to loam, full of organic matter, weeds tend to become less prolific.

You may still get weeds but the type of weed changes. For example, capeweed (*Arctotheca calendula*) and Patterson's curse (*Echium plantagineum*) are common in calcium-deficient and infertile soil, but as we build healthy soil these very unwanted plants are replaced by fat hen (lamb's quarters, *Chenopodium album*) or other "weeds" that are indicators of fertile soil.

It is quite clear that succession, for want of a better or possibly more appropriate term, occurs in soils too.

In essence, ecologically speaking, our backyards, which we enthusiastically tend and maintain, just want to grow up. So permaculture has always promoted the planting of the large climax, perennial trees and shrubs, as well as our annual vegetables.

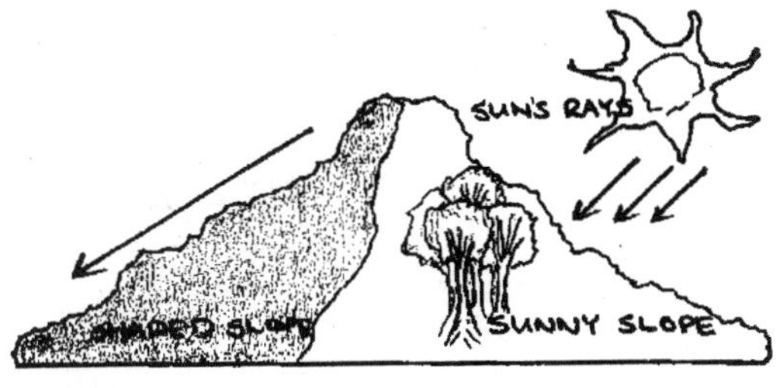

Different locations and aspects have different microclimates.

We strive to create a mature woodland or open forest, or even a rainforest in that climatic region, but what we end up doing in our backyards is more akin to a very young ecosystem that is never allowed to fully develop.

We install plant systems as a selective diversity — we determine what trees and shrubs we should have in our gardens, and in some cases this ad hoc combination may only work poorly, even with our aspiration of a "food forest" in the backyard.

Forest ecosystems have many layers and components.

As we continually harvest and remove edible plants, the many garden beds always remain ecologically unstable.

Once we understand all of this, we can implement more complex ecosystems. We can use different parts of the garden to create microclimates for the optimum growing conditions for particular plants.

Developing microclimate regions in our gardens is driven by the sun and heat. The soil heats up, light and heat are reflected, heat causes winds, and warm water evaporates and increases humidity.

In turn, plants that survive further contribute to the local environment — creating shade as they grow large, now reducing the air temperature even more, reducing evaporation from the soil, and new habitats are formed. It's like the analogy of the woodland becoming the rainforest.

Permaculturalists should all become familiar with microclimate gardening as this is the key to successful guilds (see page 23) and successful production, as well as moving the garden to maturity.

Unfortunately, our annual vegetables are fast growing and productive, and then harvested typically within a year. The ground is prepared and another crop sown.

When we have fruit and nut trees and other perennial plants, such as herbs and companions, then our gardens go through a disjointed successional phase. Perennials continue to grow and spread but annuals are replaced.

Gardening like this, unfortunately, cannot be helped if we want to grow food. It is possible, of course, to grow more perennial vegetables, but this can limit our variety of foodstuffs. Most people don't want to only eat the same vegetables all the time, so seasonal variation is preferable.

The design process

When planning a garden, it is not just a matter of drawing some garden beds and trees on a piece of graph paper. Truly functional, interactive gardens need thoughtful planning. However, as permaculture educator Toby Hemenway has said, "doing an imperfect something is better than doing a perfect nothing."

Functional gardens can be formal, but formality is a human-imposed condition. Nature is pretty informal with large numbers of different varieties all intertwined, often growing in random spacing and arrangement. I am sure nature has rules, but probably not many, so we shouldn't set ourselves a whole lot of rules and regulations when we design gardens.

We all know our climate is changing, and this may mean drying out in some areas and more flooding and rainfall in others. We need to design gardens that are resilient — that are adaptable to changing conditions, that are tough and hardy, that can be neglected and still survive and thrive — so design for catastrophe in mind. But don't despair; if you give a plant the right conditions — soil type, water, sunlight and nutrients — it will grow. So everyone can grow vegetables as these are pampered plants.

The phrase ON SPECIAL helps us focus on what we need to consider before we start developing a property.

O — Observation

Before anything is started, observe your property. You need to understand the direction winds and storms come from in different seasons and at different times of the day and night; where water flows when it rains; what insects, birds, reptiles and mammals visit or live there; where the sun rises and sets as the seasons pass; how high the sun gets in summer and winter; where the frost line is located; what soil types you have; and what weeds you get each season, which may indicate the nature of the soil or any nutrient deficiencies there may be.

Site analysis is the most important step in your design. You need to consider the way the garden is exposed to the sun. Make the most of the sun when you need it and you may need to screen it at other times.

Look at the landscapes in neighboring yards as you may want to either screen the landscape or, perhaps, "borrow" a landscape so it becomes part of your garden. Existing vegetation in your garden may be retained, or you may want to remove it if it's past its best.

Think too about the house as it's such an important part of your design. It gives you your sanctuary and living space, and, of course, it gives you the entry point to the garden.

A thorough site assessment is crucial for successful gardening. Ideally observing wind patterns, rainfall events, water flow across the landscape, erosion, temperature of the soil, sun angles and areas of seasonal shade should occur over a year or more to build a picture of your local environment.

This is not always possible as many gardeners just want to plant when they get the urge, but observation and reflection can always occur at any time. You need to be able to respond to change and adapt techniques and planting schedules accordingly.

N — Needs

A needs analysis helps us focus on what our family members want, what plants and animals want or need, and what we need to do to protect and enhance our natural environment.

Write down things like what fruit and vegetables you like to eat; what herbs you use in the kitchen or medical chest; what colors you want to see in the garden; whether you want a compost pile or an earthworm farm, or both; consider chickens or other animals in the system; and so the list goes on.

It's about growing and using abundant and healthy food — everything from herbal teas to fruit and vegetables. But what is the use of having a productive, edible landscape if you don't use what you grow?

We expect a garden to evolve, and the human element will also change as our eating habits, tastes, and food processing skills evolve to cope with the surplus and successful yields of the landscape.

S — Sectors

Sector planning mainly deals with the energies that move through the property. This includes sun angles, winds, storms, floods, fire and noise. The whole idea behind sector planning is either protection or opportunity. It considers protection from harsh sun, thunderstorms, noisy traffic, temperature extremes, frost, and natural events such as fire and flood. It grasps opportunities to harvest free natural energies (air, sunlight, water movement) to create a more resilient system.

For city dwellers, there seem to be few different sectors. You typically add visual (signs, buildings, blocking unwanted views) and social (organizations, other people, extended family). Obviously, noise (traffic, workshops, retail outlets) plays a larger role than in a rural setting.

P — Placement

Permaculture designers place every component (see Elements next) in a region called a zone. Zones are imaginary lines around a house and property.

There are five zones, and all elements are placed according to how often we need to visit — with vegetables and herbs, which we use every day, in zone 1 closest to the house, and fruit trees, which we visit seasonally, further out in zone 2 or 3.

E — Elements

No, we are not talking about the chemical elements you learned about in school. Everything we place in a design is called an element, so this includes plants, sheds, garden beds, ponds, trellises, chicken pen, compost pile and so on. And don't forget about the entertainment area, the gazebo or pergola and barbecue.

We endeavor to match and group elements together so that the products of one element may become the needs of another.

For example, the manure from chickens is used to feed the earthworms or add to the compost heap, so it makes sense to have the worm farm and compost area close to the chicken pen.

While it is important to grow as much food as possible, don't think that this is all you should grow. Recognize that some plants are grown to help with pest control, a shady tree to sit under, others because they smell nice

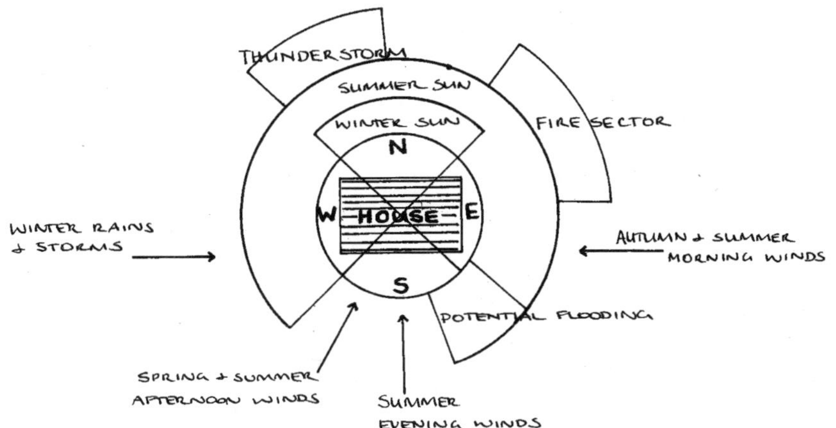

A sector plan highlights the movements of energies through the system.

and look attractive, some to protect sensitive plants (nursery trees) and some to attract predators or pollinators.

In many ecosystems every plant has a role to play, so when we design a garden try to mirror these ideas. It's all about balance.

C — Collaboration

Every plant and animal likes company, and some combinations and relationships seem to benefit all involved.

This assembly of organisms and parts of their habitat or collections of interacting, networking components is called a guild.

Common companion planting where combinations of vegetables and herbs are planted close to each other is a good example of a guild.

We're not too sure about how some guilds work, but we recognize that many plants struggle to get going. Having a nurse or chaperone plant nearby to protect vulnerable seedlings and to create ideal growing conditions for the seedlings will ensure greater survival.

We just need to work out the right guilds — the right relationships and combinations of plants and elements that are mutually beneficial. The classic example of a guild is the famous "three sisters" of Native Americans:

- Corn — is the trellis (growing frame) for beans. This is planted first and when the corn seedlings are about 6 inch high the other seeds are planted, sometimes together, sometimes separately.

- Beans — are nitrogen-fixing. Bean seeds are often planted later than squash as they are fast growing and they need corn to become established before they start climbing.
- Squash — shades the ground and suppresses weeds.

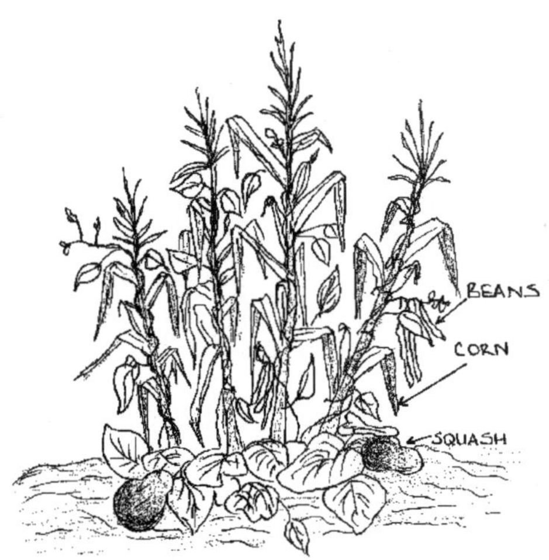

Example of companions — the "three sisters" of corn, squash and beans.

In North America, a fourth "sister" is often added, the bee plant (*Cleome serrulata*), which attracts beneficial insects for pollination (for beans and squash, as wind pollinates corn) and has edible leaves and flowers. The Incas from South America also added amaranth, which is a grain and useful dye.

In other parts of the world common herbs such as borage (attracts predatory wasps and bees for pollination, edible flowers) or perennial basil (butterfly and bee pollination, edible leaves) substitute for bee plant.

There are many other good examples of guilds and companions, and these combinations of elements are a crucial design idea that we all need to adopt.

So, let's focus on implementing noncompetitive guilds and resource sharing when we endeavor to develop functioning and beneficial gardens. For example, we should underplant fruit trees, such as apples and lemons, with comfrey, borage, fennel, yarrow, and non-climbing (shrub-like) beans or peas. Nasturtiums can be used as ground cover, but so can nitrogen-fixing perennial clovers and alfalfa. It's probably not so wise to have tall plants under our fruit trees.

The main concept here is that not all plants in a garden have to be edible. We need a mixture of plants in this artificial guild to provide different functions: grass suppression; food for insect, bird or mammal pollinators; nutrient accumulation; mulching plants; nitrogen fixers; pest repellents; soil fumigants (nasturtium, marigolds); and soil building.

We also have plants to create habitat, attract and protect large predators, and provide shade and shelter. For example, yarrow and comfrey are high in minerals and have medicinal functions, while borage and fennel attract pollinators and predators, and both have edible parts.

While garden guilds are designed by humans, you wouldn't put all of these plants under a lemon tree. It's a deliberate assembly of plants that we choose to grow in the belief that we are mimicking a community in nature.

Even so, in disturbed bush or forest areas we often see exotics amongst native plants. Interestingly, they all seem to flourish, and this suggests that nature uses whatever plants it requires to fill the niches in that ecosystem. Nature doesn't distinguish between natives and exotics, unlike some humans who fixate on exotic removals in the belief that nature will suffer if we don't.

Under the umbrella of "companions" we need to include pest control, and design for scented walkways, pest-repellent herbs and predator attracting plants (tansy for ladybugs, yarrow and fennel for hoverflies and wasps), as well as plants to attract bees and butterflies for pollination.

Generally, beneficial insects are either predators, parasites (parasitoids), pollinators or weed feeders, but let's focus on predators.

You always need to have a few pests around to encourage predators so that the response time between pest attack and predator dining is reduced. You should use host plants to always maintain predator species in the garden, and these include buckwheat, coriander, clover, fennel, lemon balm, parsley, Queen Anne's lace, yarrow and tansy.

As part of this strategy it is also important to leave some herbs and vegetables to flower and set seed. Besides the obvious idea of saving your seed, this encourages predator adults, such as hoverflies and lacewings, to feed on the nectar and pollen and lay their eggs on the plants. It is often the larvae of the species that are the real predators and these devour aphids and other sap-sucking pests as they grow.

I — Irrigation

Water is essential to all life, and if we want to grow plants or keep animals we need water. When planning a garden think about how the plants will be irrigated. You might choose to hand-water, but ideally having an automatic watering system is best. Dripline, individual drippers and water-efficient sprinklers are all suitable and some of these components are discussed later in chapter 10.

Think about how you could include and incorporate stormwater, rainwater and graywater as strategies to water gardens and the landscape.

A — Assets and aesthetics

Assets are the resources available and materials required for the project. Resources can be:

- Living — plants and animals. What you have already, what you can easily source, what may be available at the local nursery.
- Structures — building materials, rocks and timber. Is there timber that can be milled on-site, rocks that can be used for retaining walls, or a local farmer who has straw for the walls of buildings?
- Energy — wind, wood, water (and machinery to do work). Are there opportunities to harvest or generate electricity or other forms of energy in some way?
- Social — people and their skills, suppliers and consultants. Who is a local "expert" you can call on to help with plant selection or breeds of chickens or grafting techniques?
- Financial — what money is available to purchase labor, machinery or materials? Can some materials or plants be bought with LETS (Local Employment [or Exchange] Trading System) and other alternative schemes? Is there potential income from the sale of goods and services?

Professional landscapers will often produce a bill of quantities: an itemized list of what is required for the project, with specifications for some components, and often anticipated costs so that a budget can be determined.

Aesthetics is about beauty, so think about flowers, foliage, color, and the visual setting (pretty view).

L — Landscape

This considers soil, topography, aspect and slope. Paths and access tracks or roads can be listed here too. Topography refers to terrain — how rough, steep or slippery, or whether it is rock-laden, a gravel pit or just sand. Aspect is about orientation to the sun (and direction). Some slope on a property is good, but there would be some instances when the slope is too steep for vehicles, houses and any cultured ecologies.

As mentioned already, soil is discussed in detail in the next chapter, as this is the key to ecosystem yield and growing healthy and safe food.

Designing for others

Designers need to ensure that gardens and places connect with the inhabitants and engage them. If this doesn't occur they won't stay in the garden.

The designer's mantra is that we're in it together, helping each other. Everyone counts, everyone has a say, so that everyone has ownership. It is not just the designer but all family members have to be involved.

Producing a design or plan is generally undertaken in a number of steps, involving the client in some stages, and just the designer or design team for the rest. Think of the acronym SPEER.

S — Species selection

Further to what we discussed earlier, particular species should be selected with climate, growing conditions, soil type, temperature extremes and growth habit all in mind. While a client may desire a particular fruit or vegetable, and maybe we could obtain them from a nursery, you may find that some plants just won't grow in that location.

P — Planning

Good planning involves visioning and patterning. You need to be able to imagine what a garden or property will look like as you undertake a design. The "design" is not only artwork, but it is also a representation of a functioning cultivated ecology. Permaculture designs can be full of patterns. Patterns are all around us in the shapes of honeycomb, the arrangement of seeds in a flower, the spirals of galaxies, the flow of water across a landscape, the eddies of winds, and the crystals of water in snowflakes.

Some permaculture designers might acknowledge the patterns in nature by drawing in mandala gardens, herb spirals and circle gardens; but none of these, I might add, are necessary in any permaculture design.

Planning a garden also requires commitment. You need to look after all living things and make the time to ensure they are watered and fed, pruned and mulched — even transplanted to another area if they appear to be suffering — and pests and diseases controlled. I don't think gardeners really appreciate the amount of time that is required to ensure their little ecosystem is humming along just nicely.

Any garden plan or design has to be done to scale. The usual way to do this is on graph paper (A3 at least) and have lots of cutouts and shapes of garden beds, structures (sculpture, artwork, trellises, sheds), trees and anything else you want in the garden (playhouse, barbecue, rainwater tank, scented garden).

Garden beds can be of any shape. Square or rectangle beds give the impression of sharpness and formality, rounded or curved beds seem softer and send an invitation to meander through the garden. At the end of the day, it's about how you feel in a garden or space, and what combinations of elements seem to work.

Over time gardens change. You put in a trampoline or small pool, then remove these when the kids have grown up and left home, a tree once planted for shade now takes up the whole backyard and must be removed, or some fruit trees and other plants reach the end of their lifespan and die. When you are designing in those early stages try to plan for the future. Obviously who knows what the future holds and what changes you want to make in the garden, but at least plan for access and contingency. Leave a path wide enough for a small machine to come and help with plant removal or bringing in soil and mulch, or to enable a rainwater tank to be rolled around to the back. Don't fill every space unless you are happy to remove any plants to enable access or you are sure that what you are building will last forever.

E — Evaluation

Once an initial draft has been done (artwork and report), then this is shown and discussed with the client who will be able to provide some feedback about what they like and what they would like to see changed.

This enables you to make amendments and work towards the final design and report. Of course, what is finally produced will, no doubt, pass through many stages and many meetings with clients. Ultimately, there may be tradeoffs between the client wants and the needs of the other organisms in the ecosystem and the environment itself.

E — Execute

Executing the plan is implementing the design. It is rare that a whole design is built at once, and the most common implementation is spread over months and, more usually, years. This is where you need to set a schedule of works, with realistic goals and a workable budget.

R — Reflection

This is really reevaluation, where changes to the design are made before building begins or changes made during implementation are recorded.

It allows you, as a designer, to critically examine what seemed to work, what needed to be changed and what should be discarded. Too often designers put in particular plants, structures and garden bed shapes, simply because they want them included or in the belief that as these things worked before they should work again here.

The design always has to be what is best for the clients, the site and the environment.

Implementation schedule

Once you have a design for the property the next step is building it. Think of the acronym PERFECT, where each letter indicates an important part of this process.

P — Priorities

This includes buying or obtaining their personal choice of fruit trees or vegetables, and what you want to plant first. It is also about pre-lays for irrigation, marking out paths and structures, and building some of these as required.

Not everyone has a big backyard to grow their own food, and not everyone lives in a passive solar home. So if you have some garden area you may be faced with a choice — do I grow some food or do I plant to cool my house. Well sometimes you can do both, using deciduous fruit trees for summer shade and food and climbers, like kiwifruit, cucumbers and pumpkins, strung over a trellis or some structure to protect walls from the hot summer sun.

While there is great joy and satisfaction when you grow some of your own food, it may be better to focus on being comfortable in your home and reducing energy bills for heating and cooling.

E — Environment

Our aim should be to build soil and to repair the land. One of the ethics of permaculture is "care of the Earth" and this has always been the foundation on which the philosophy and principles are based.

R — Resources

These are the materials available for use, to build structures (trellises, propagation tunnels, sheds, retaining walls), to fill garden beds (compost, soil, mulch) and to plant (what herbs, trees and shrubs you have assembled already or are willing to buy). Using, accessing and assembling resources is an art in itself.

F — Finances

Most things you want to implement cost money. You should determine a budget, but allow for contingencies, as often projects seem to take longer and cost more than you had set aside. While you can scavenge lots of materials from roadside pickups, neighbors and friends, and obtain many things for free or for "green" dollars through LETS and freecycle networks, you may still have to pay for contractors, machinery and some building materials.

E — Earthworks

These are the landscape changes that are undertaken and include drainage, mounding garden beds, putting in a driveway or access track, running power to a rear shed, burying reticulation pipes, clearing an area of vegetation, leveling a site, shaping a pond and laying hardscaping (hard surfaces such as roads and parking areas). Earthworks are normally undertaken early in a project, and if you have a machine available, use it.

Machinery is often used for earthworks — digging trenches, moving soil and so on.

C — Consideration

These are mainly seasonal issues, such as availability of bare root fruit trees, planting schedule (best when rain season is about to come) and weather at the time (who wants to work in the rain, stand in the hot sun all day and go out on a frosty morning?). You need to also consider the availability of electricity to power tools, water to mix concrete or soak plants, and access (do you need a four-wheel-drive vehicle to get to the rear of the block?).

T — Technical

Sometimes you need to rely on tradespeople and contractors to undertake trenching for the pre-lay of irrigation pipes or running power cable and conduit for pumps, a licensed plumber to connect your graywater system, an electrician to install a power point near the pond to operate the fountain pump, a concrete finisher to float the floor of the new shed, and a stonemason to build a rock wall or limestone block wall.

At other times, you just need advice on the schedule of works, where to place a rainwater tank, what size pump you would require to irrigate the gardens, and what trees should be removed from the property to have good solar access.

» DID YOU KNOW?

Many deciduous fruit trees can be ordered and collected as "bare root" plants. This is because these plants are dormant in winter, and can be literally dug out of the ground, transported and then replanted with great success.

Sharing the surplus, thriving and surviving

Here are some hints about getting a garden up and running and to make it something special:

- In small urban lots it is impossible to grow lots of fruit trees. However, if you become friendly with your neighbors, you might find them happy to share what they have and you plant what is missing to supply them in return. Your neighborhood becomes your orchard.
- Plant a number of native plants (preferably indigenous to your area) as these will attract wildlife: nectar-producing species to attract butterflies, insects and small birds; grasses for reptiles (lizards, skinks and geckoes), which will help with pest control; and emergent macrophytes (reeds and rushes) near a pond for frogs.
- Provide hollows, logs, piles of rocks and sunbathing spots for lizards, and a garden pond for frog habitat. A water supply is crucial to attract wildlife — water is a life magnet.
- Install a few nesting boxes in the taller trees on your property. Many birds and animals that use nesting boxes exist on a diet of common insect pests.

  Place nesting boxes as high as possible — away from predators. Make sure no direct sunlight or rain can enter. Monitor the box to ensure that bees and feral birds do not take up residence. Secure firmly to the tree.

Provide nesting boxes to encourage wildlife to visit.

Building plans for boxes are easily obtainable, as the entry hole needs to be a particular size for each bird or mammal species. Never paint the inside of the box and only use nontoxic paint on the outside.

However, there is no guarantee that nesting boxes will be used by the desired species, but having birds and mammals to view is enriching for the soul.

- Build some garden beds on themes. These include a pizza garden (herbs used to garnish and flavor a pizza); the scented garden; a senses garden — to see, hear and feel (plants based on smell, texture, color and sound, e.g. leaves rustling in a breeze); and play — children having their own vegetable garden, trees they can climb or plants to attract butterflies.
- Borrow a view. Wouldn't it be great, sometimes, if we didn't have fences between properties?

  Your garden would blend into your neighbors. The large majestic tree next door would become a backdrop for your understory garden nearby. When planning your garden use the views and features of adjoining properties to extend your own.

- When laying out the garden areas, use spray paint, string, small rocks or pavers to mark out beds and paths. This enables you to have a good look at what the garden area will look like, and allow you to make any changes before all the fun begins.

Borrow a view from a neighbor's garden.

- Active children need to be kept active. They can collect eggs, feed earthworms, climb a tree or up into a treehouse, swing, play in a sandpit, or throw balls through a basketball hoop.
- If there are no children nor any prospect of children in the house, some other household members can perform some of these tasks such as feeding chickens and earthworms.
- Take over the curbsides! What an under-used space in most suburbs.

GETTING THE SOIL RIGHT 4

What is soil?

SOIL IS A LOT MORE THAN SAND and some leaf litter. It is a complex mixture of physical, chemical and biological components, in various combinations of these to create a wide range of different soils.

While it generally contains three different-sized particles (sand, silt and clay), the amounts of organic matter, water and soil organisms also vary, and all of these can be manipulated to produce a custom-made soil.

Plants typically grow in soil. Every soil type from sand to clay has some plants that can grow in it, but we tend to grow our edible crops in a mixture of soil particles we call loam. Loam is that mixture of clay, silt and sand that seems to have the best properties for growing vegetables — enough clay to hold water, enough sand to prevent waterlogging and enough spaces for air. Add some organic matter (up to 5% is adequate) and you have the perfect soil for growing food.

Loam contains about 40% sand, 40% silt and 20% clay, and on a small scale in a backyard garden bed these ratios can be changed. We can easily change soil by adding bentonite (clay) to sand, compost to nutrient-deficient soils and gypsum to break up some types of clay.

Nutrient levels in soils change all the time. It is the nature of plants that they decrease soil fertility, so we need to develop and use strategies to replace lost nutrients and fertility.

This is why many farmers use the "ley" system of resting a paddock and sowing it with clovers or other legumes, which repair and build soil again ready for the cereal crop next season.

This is why backyard gardeners must add compost, manures and other amendments to rebuild the soil ready for next season's food crop, and why it is important to also rotate crops so that different types of plants can re-aerate the soil and re-introduce

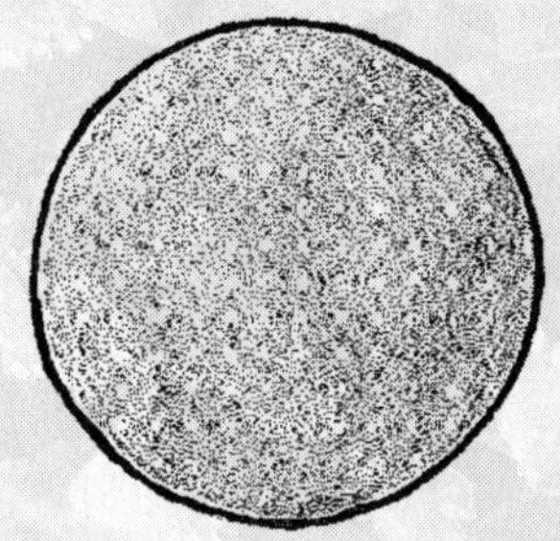

Soil particle sizes. Smallest to largest — clay, silt, fine sand, coarse sand. The largest particle in reality is 1⁄16 in diameter.

organic matter to build healthy soil once again. As nutrients get removed by the plants we harvest, they must be replaced.

Let's examine some of the armory you can have to create living soil, which, in turn, will create healthy food and ultimately healthy people. Soil health and soil quality are the keys to sustainable agriculture.

Compost making

The best compost is made when it is a "batch" process. This means that the materials are assembled and the compost pile activated. With proper preparation, piles can be made to heat and decompose without the need for bins. This is different from what most people do — they start a small pile, typically in a commercially made plastic bin, and periodically add to it. The pile never gets large enough or hot enough to break everything down.

A compost bin with removable sides is the best option. Each side should be about 3 ft to 4 ft wide and high. This allows you to access the compost from more than one direction, and permits easier shifting from one bin to another. This saves using the pitchfork to lift the compost high over a bin side.

When choosing a bin, always remember that good air flow and the size of the bin, not what it is made from, determines compost production. The more open the bin, the more air can enter the pile. Avoid using the wall of a building or a fence structure as part of the compost bin or pile. Active composting will decompose walls, timber and paint.

A rotating barrel is another option, but these are limited as they generally only hold a small amount, and larger barrels are just too heavy to turn.

Composting can follow one of two pathways: cold composting and hot composting.

Cold compost

Some people prefer to exert themselves less and let time, nature (and earthworms) do most of the work. Cold composting is a more passive, gentle approach.

The volume is not critical, as the pile never heats up. So you can have a series of bins, and when one bin is full, starting filling the next. A technique that is adopted by many organic growers is sheet mulching, and this is essentially a cold compost made *in situ*. Organic garden material is broken down in the bed itself, rather than accumulating and transporting the materials to a dedicated compost bin or pile.

The cold compost pile can have problems, and may attract animal and insect pests, including mice, snails and wood lice.

One smaller-scale cold compost strategy is the bokashi method. Here your kitchen food scraps are placed in a bin and a small amount of granules, inoculated with microbes, is added. The mixture is compressed and the bin sealed with a lid, making the process anaerobic. It is more akin to fermentation where foods are "pickled". Using fermentation to make fertilizers is discussed later in this chapter and to make or preserve foods is discussed in more detail in chapter 14.

Hot compost

This is by far the best method, but it does have its secrets. You need to follow the six Ms: Materials, Moisture, Mixing, Microorganisms, Minerals and Mass.

1. Materials: the "right stuff"

For best results, a compost pile must be, as the word implies, a composite of different materials — a mixture of plant and animal material.

To make everything work properly you need a balance of carbon and nitrogen substances. Carbon substances are "brown", and these include plant materials and sawdust, straw, paper, cardboard and dried leaves.

Nitrogen substances are "green" and contain protein substances, so again living plants and weeds, but also animal manures and food scraps.

The ideal carbon to nitrogen (C:N) ratio is 30:1, where decomposition organisms require the carbohydrates and carbon substances to be balanced by a suitable proportion of protein or nitrogen.

Most deciduous leaves have a C:N ratio of about 60:1, while grass clippings, manures and food scraps have a ratio of about 20:1 or less, and woody materials often range as high as 500:1. Too much nitrogen in a pile results in the formation of ammonia gas; too much carbon and the pile will take years to break down.

In a practical sense, about two shovelfuls of fresh animal manure to every wheelbarrow load of green plant material, such as weeds and lawn clippings, seems to work well. If you include some paper or dried leaves, add more

Material	C:N ratio (weight:weight)
lawn clippings	20:1
weeds	19:1
paper	170:1
food wastes	15:1
sawdust	450:1
chicken manure	7:1
straw	100:1
seaweed	25:1
cattle manure	12:1

Table 4.1. C:N ratio of common materials.

animal manure. If you cannot access animal manures you can use blood and bone fertilizer or fish meal as your source of nitrogen.

2. Moisture: add water

You need to add enough water so that when you pick some material up it is like a damp sponge, and when you squeeze the material a few drops of water drip out.

About 50% of the pile should be water (you will be surprised how much water is required!). If there is not enough water (<40%), then little decomposition occurs, but too much (>60%), then less air is available in the heap and it becomes anaerobic (without oxygen). Under-watering is the largest single cause of slow composting.

3. Mixing: the air that I breathe

Generally, the more oxygen then the more heat, quicker decomposition, and less smell (and less flies). We want to make the pile aerobic (with oxygen) to enable the pile to really heat up. Anaerobic composting, with very little air in the pile, causes problems such as smells often associated with decomposition (think septic tank).

So, turn the pile as often as you can — once a week for the first three weeks and then leave alone. Turning with less frequency will also result in a good compost product, but will necessarily take longer. However, turning compost too much can waste valuable nutrients into the air and cause the pile to lose too much heat.

Getting enough oxygen is crucial to the success of the hot compost process. A composting grate, at the base of the pile, is a useful strategy to help with this. Placing small stones, twigs or other coarse material on the bottom of the pile or bin allows air to passively move through the pile, providing microorganisms with the oxygen they need to enhance the decay process.

4. Microorganisms: the engine room

You don't have to source the special "brew" of compost microorganisms (mainly bacteria) as these are always present in the air and surroundings.

When you start a new compost pile, you can add some "aged" or cured compost from a previous effort. This inoculates the mixed materials with bacteria to "kick-start" the new pile.

While you can buy commercial inoculants, it is far cheaper to use aged compost, dilute urine or herbs such as yarrow, comfrey and borage.

It is believed that these herbs contain high levels of nutrients to help feed the microorganisms, which enables them to build up in high numbers very quickly, but little research has been undertaken on this. It's up to you to investigate by trial and error.

5. Minerals: is that sand you're adding?

Plants need soil. They need minerals in soil. You should never just grow plants in pure compost. I know you can, and I have many times, but a soil with only 5% organic matter is more than ample to support plant growth. Compost is expensive so it makes sense to only use it mixed with sand.

While you can just use sand, you should experiment with what you add to the compost pile. Small handfuls of rock or granite dust, loamy soil, crushed limestone and diatomaceous earth all add nutrients to the pile and these are beneficial to growing plants.

Adding sand also helps earthworms — they use the sand grains in their gut to grind dead matter. But be warned: adding lime or too much limestone does more harm than good.

Even though these substances add calcium to the compost, it is not required for composting to proceed.

There is also no need to "neutralize" the acidic nature of the compost as it decays, as well-made compost goes through an acidic stage before it finally balances itself.

6. Mass: size matters

There are two aspects to this — the amount (volume) and particle size. The shredding of materials is important. The finer the organic material the faster the decomposition (large surface area to volume ratio (SA:Vol) so microbes can attack all sides).

While it is best to shred the material you use in the compost pile, occasional bulky, woody material, such as wood chips and street-tree prunings, assist in aeration of the pile.

Also a rounded pile has a low SA:Vol ratio so less heat is lost, whereas a long, thin pile has a large SA:Vol ratio and will cool down quickly.

Secondly, piles require a certain critical mass — you need at least 35 ft^3 but larger is better. This is a big pile. If the pile is any smaller, it cannot maintain heat, heat escapes and the pile cools down.

Large piles contain the heat longer. This results in better pasteurization and the microbes have more time to degrade any toxins that may come with the raw materials.

You might have to collect enough material over a few weeks before you activate the pile to start the hot decomposition process. The dedicated organic gardener will ask neighbors for their plant prunings and lawn clippings.

Making a compost too big, say 50 to 70 ft^3, makes turning and shifting the pile difficult and laborious. There is also a natural tendency to continually add material as you acquire it. The pile slowly builds up. However, this will usually result in passive or cold composting and rarely does the pile heat up or maintain the heat to sustain the compost process.

Troubleshooting your compost making

Some of the problems you may encounter and their solutions are listed opposite.

However, when a compost pile is hot (140°F) and active for several days, it is possible to add small amounts of additional material such as food scraps.

Adding food scraps to the compost pile may be necessary if you have food scraps to deal with, but it is probably better to bury the scraps in the garden and make a new pile whenever you have assembled enough material.

Another reason why it is important to start with a big pile is that during decay the pile shrinks, and as it shrinks it loses heat to the surroundings.

You can monitor the heat of your compost pile by using a compost thermometer. These are long versions of cooking thermometers.

A thermometer takes the guesswork out of when to turn the pile, when different stages of the composting process are occurring and when it is cool enough to use on the garden.

After the compost pile cools down (typically in a month or two), then other creatures invade. These can include earthworms and a host of small invertebrates, such as insects.

Not all insects in a compost pile are "pests," as the compost ecosystem includes a host of useful invertebrates, including isopods, millipedes, centipedes, worms, and ants among others.

Identifying organisms with a hand lens or microscope is a useful strategy to determine the health of your compost or soil.

Having a few compost bays permits decomposition at various stages.

Problem/issue	Causes/reasons	Remedial action
Bad odor	Anaerobic pile	Turn materials to aerate. If too wet, check for proper drainage and mix in dry leaves or straw.
Insect pests	Too dry, not mixed properly	Make sure that if you use food materials they are properly buried in the center of the pile. Use caution if termites are in area — don't use wood chips, cardboard or newspaper in the pile.
Pile not breaking down	Insufficient nitrogen Pile is too dry Poor aeration	Add grass clippings, manure or some other natural nitrogen source. Add water while turning, until moist. Start turning and mixing materials more often.
Pile heats up, then stops	Poor aeration	Hot piles need lots of fresh oxygen so turn materials as pile starts to cool down. It might be necessary to add a nitrogen source such as animal manure too.
Weeds growing out of the pile enough	Pile is too dry, and certainly not hot	Usually add lots of water to get the right amount to kick-start the composting process.

Table 4.2. How to solve some of those compost making problems.

Soil amendments

Soil amendments are substances that you add to the soil to change its nature. Broadly, there are two categories: chemical-based and biological-based.

Chemical-based amendments

Chemical-based amendments are natural rock minerals, chemical salts or manufactured substances readily bought or obtained.

Some of these chemicals are mined and available in their raw state, others are manufactured as by-products of industry or from other ingredients. For example, natural gas can be changed into ammonia, which is then changed again into ammonium nitrate, a fertilizer commonly used on large-scale farms.

Table 4.3 lists some chemical-based amendments and how they can be used.

Common name	Chemical name	Uses	Application rate
Gypsum	Calcium sulfate	Breaks down clay, provides calcium and sulfur fertilizer.	Gypsum should be added at a rate of 10–15 lb/100 ft^2, digging into the soil.
Bentonite (clay)	Hydrated sodium calcium aluminum silicate	Increases the soil's ability to hold water and organic matter, predominately in sandy soils. This clay also has the natural property of polar attraction, so inherently will draw water and nutrients towards it, which reduces hydrophobic conditions.	About 5 lb/100 ft^2 (more in very sandy, water-repellent soils) and work into the top 4 in of soil, and water in well.
Sulfur	Pure element (symbol S), yellow crystalline solid that is insoluble in water.	Dusting sulfur is a fine powder fungicide. Powdered sulfur is the coarser grade used to lower pH in soil gradually (makes more acidic).	To lower the pH, add 0.5 lb/100 ft^2 for sandy soils and up to 1.3 lb/100 ft^2 for clay soils. Start with this. Add more at the same rate, a few weeks later, if pH is not down enough.
Rock dust	Ground rock (e.g. granite, basalt) from quarries. Can be a fine dust or coarser-grained for ease of use. Contains wide range of elements, both macro- and micronutrients.	Replaces minerals in leached, poor quality or overworked soils. Used to boost and aid composting.	Apply on the basis of little and regularly. 1 lb/100 ft^2 once or twice a year. 2 lb (four handfuls) per tree at planting.

Common name	Chemical name	Uses	Application rate
Spongelite	Silicon dioxide (silica). From fossilized sponges.	Water retention in potting mixes and garden beds.	General rule: the poorer the soil the more spongelite is required. Typical rate between 4–20 lb/100 ft^2.
Dolomite	Calcium and magnesium carbonates.	Increases soil pH (counteracts acidity). Source of calcium and magnesium for plants.	Garden beds, from 1.5 lb/100 ft^2 (sandy soils) to 10 lb/100 ft^2 (heavy soils).
Limestone and lime	Calcium carbonate Calcium oxide	Raising pH, source of calcium. Limestone more gentle than lime.	About a handful per 10 ft^2. Sprinkle on surface and allow rain to wash into soil.
Zeolite	Different forms of crystalline aluminosilicate. Either a natural substance formed from volcanic ash or can be artificially produced.	Can hold and exchange nutrients required by plants, making nutrients readily available.	It can be worked into the top 4–8 in of soil, or added to compost piles. Use at a rate of: sand 5–10 lb/100 ft^2 clay 1–2 lb/100 ft^2 compost 20 lb/100 ft^3.
Potash	General name for potassium-based salts including sulfate, nitrate, carbonate and chloride.	Adds potassium. All of the potassium salts can be used as fertilizer.	Apply at a rate of 1 lb/100 ft^2 and water in thoroughly (potash salts are generally water-soluble).
Rock phosphate	Calcium phosphate	Naturally mined mineral that supplies phosphorus, a major plant nutrient.	Half a handful per 10 ft^2. A slow-release fertilizer.

Table 4.3. Chemical or mineral-based soil amendments.

Troubleshooting your soil

Table 4.4 lists the types of amendments to use to change particular soils.

Soil problem	Amendment
Too much clay	Gypsum
Too much sand	Bentonite
Acidic	Limestone, lime or dolomite
Alkaline	Sulfur
Poor water retention	Zeolite, spongelite or bentonite
Poor nutrition — little nutrients	Rock dust, zeolite, potash, rock phosphate

Table 4.4. What amendments to use for difficult soils.

Biological-based amendments

Biological-based amendments are those that are produced from once-living things. These substances can be made from dead and decaying organisms, burning organic matter and from the metabolic wastes of animals.

Organic soil amendments

Organic-based materials certainly improve soil structure, as well as providing nutrients to plants and microorganisms in the soil, and Table 4.5 briefly lists some of the different types.

Common name	Chemical name	Uses	Application rate
Compost	Decayed plant and animal residues.	General plant fertilizer. Increases organic matter and humus in soils.	About two shovelfuls to a wheelbarrow of soil.
Wood ash	Ashes from wood burning contain potassium, calcium, phosphorus and magnesium.	8% potash — potassium compounds. Reduces acidity, repels pests.	Has about 50% the liming ability of lime (calcium oxide). For general use sprinkle 8 oz/100 ft^2.
Charcoal, biochar (Carbon)	Made from burning plant matter in little or no oxygen at a high temperature.	Source of carbon for micro-organisms in soil.	A shovelful every 10 ft^2 of garden bed.
Mulch	Wood pieces, shredded plant material — leaves, stems, roots.	Surface cover to reduce water loss from soil, prevents over-heating. Some nutrients available to plants when broken down.	2–4 in cover over soil and around plants.
Urea and urine	Urea (carbamide) has chemical formula $CO(NH_2)_2$. Contains 46% nitrogen.	Good source of nitrogen. Urine is slightly acidic so can be used to ameliorate alkaline soils.	Dilute one part urine to 10 parts water. Urine is not sterile so only use on the soil, not onto crops.
Blood and bone	Waste products from abattoirs. High in nitrogen, calcium and phosphorus.	Slow-release fertilizer, providing gentle long-term feeding of plants and soil.	1–2 lb/100 ft^2.
Peat	Compressed, partially decayed plant material, high in carbon.	Source of carbon. Is slightly acidic so good for blueberries and to change alkalinity of soil.	For acid-loving plants use 30 to 70% of peat mixed with soil.
Sawdust	Wood crumbs from sawn timber.	Used to make paths in garden (doesn't break down too quickly). Barrier to slugs and snails.	4 in thick for paths.
Hugelkultur	Whole logs buried, creating a mound or raised garden bed.	Great if you have large tree branches you can't convert to compost or mulch.	Build rotting wood pile 1.5–3 ft high.

Common name	Chemical name	Uses	Application rate
Pine needles	Traces of nitrogen, potassium, phosphate, calcium and magnesium found in tests conducted on pine needles.	Gradually lower soil pH as these are acidic. Adds organic matter to soil. Prevents growth of some weeds.	Mulch at 4 in thick for general use. Large amounts are needed to lower pH.

Table 4.5. Organic-based soil amendments.

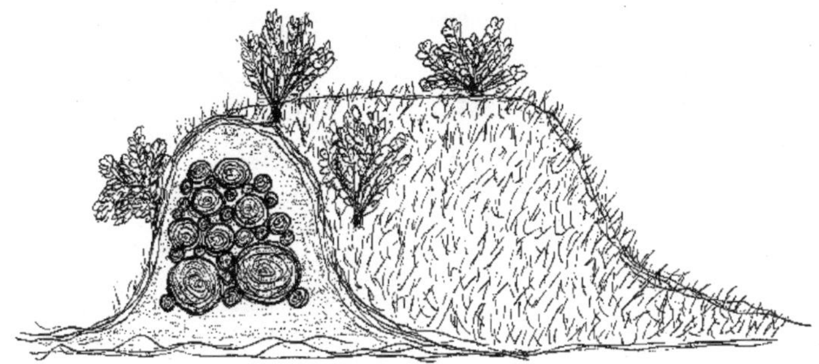

Hugelkultur — burying logs in mounds.

Mulch

Mulch is something that covers the soil. It is generally the product from shredded plant material, but it also includes other non-plant materials.

Coarse mulches are best for the garden as they allow rain to filter through into the soil below.

Mulch can be either dead material (both organic and inorganic) or a living mulch, and both have advantages and disadvantages.

Organic mulches

Organic mulches are produced from organisms, and these are usually plant-based. Animal manure can be used as a mulch, but we think of manures more as fertilizer. Straw, hay, grass clippings, shredded plant materials (e.g. leaves, bark) and pine needles are the most common mulches used in gardens. Organic mulches eventually break down and become nutrients for plants.

Mulch around a tree.

Mulch is great on garden beds, most of the time. It's not so great in winter, spring or autumn in areas that suffer frost. The mulch layer prevents the heat held in the soil from escaping (much like a blanket keeping us warm at night) and keeping the air above the ground warm. So frost settles on the mulch and the nearby plants, killing young plants and seedlings and often severely affecting many larger established plants too.

Inorganic mulches

These include natural materials such as stone, gravel and scoria (a volcanic rock), or man made (synthetic) materials such as plastic sheeting (both clear and black), geotextile and vermiculite. Table 4.6 highlights just some of the mulches you can use to protect the soil and conserve water.

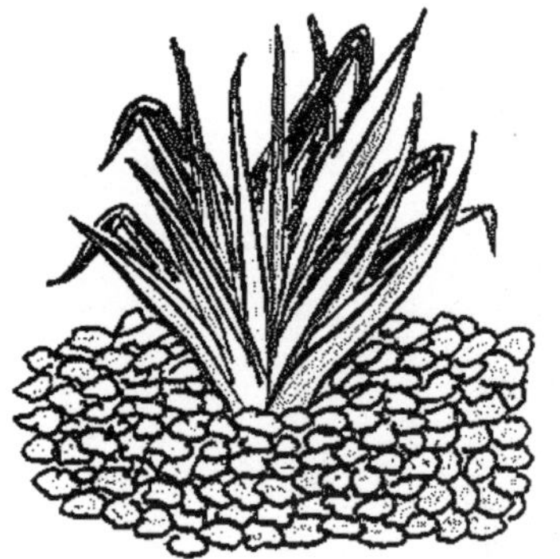

A stone mulch still provides protection for a plant.

Living mulches

A living mulch is a plant that is basically a ground cover. It spreads over an area covering the soil and minimizing the risk of erosion. Usually they are fast growing, thick and hardy; this enables them to smother and out-compete weeds.

Some living mulches are planted in a bare area, but others are planted alongside or underneath the main crop. When this occurs, the living mulch may have to be mechanically or chemically killed to enable the main crop to thrive. It is important to manage the competitive relationship between the living mulch plant and the main crop.

The examples of living mulches that follow are generic. Every country has its own plants that are commonly used for this purpose, and there are many plants that become weeds in different environments and different climates and soils. I have found, for example, vetch to be a vigorous ground cover that grows so fast in my soil and climate that it out-competes everything around it for light, nutrients and water. It produces prolific seed so it has spread everywhere on the property. It has become an unwanted plant, a weed. In other soils and climates it may be subdued and manageable.

Sweet potato is an edible living mulch.

Many living mulches are perennials, but some are annuals, and these latter types are best used when you only want a temporary cover while your main crop gets established.

Some that are listed here are herbs, and these tend to be underrated by gardeners. Many herbs are great ground covers and are sun-hardy and tough.

Mulch type	Notes
Hay	Cut cereal crops or long grass. Often contains seeds, which tend to sprout once you water or it rains.
Straw	Cereal stalks (seedhead harvested), little seed.
Grass clippings	Temporary mulch. Dries out quickly if thinly applied.
Street prunings	Large plant fragments, coarse, relatively cheap, ideal mulch.
Stone, gravel, scoria	Useful when no plant material available.
Plastic sheeting	Clear plastic film can sometimes be useful to solarize turf and weeds.

Table 4.6. Common types of mulches.

Living mulch	Notes
Clover *Trifolium* spp.*	Main varieties white and red. N-fixing**, some are annuals, others short-lived perennials, forage.
Vetch *Vicia* spp.	Annual, N-fixing, forage.
Kidney weed *Dichondra repens*	Useful, perennial lawn, but prefers semi-shade, moist areas.
Pigface *Carpobrotus* spp.	Succulent, hardy plant that tolerates full sun, poor soils and windy conditions. Salt-tolerant, edible fruit.
Sweet potato *Ipomoea batatas*	Main types brown and white fleshed (and occasional purple). Besides edible tubers, the leaves are also edible. More nutrition than common potatoes (*Solanum* sp).
Nasturtium *Tropaeolum majus*	Spread easily. Edible leaves and flowers.
Fescue *Festuca* spp.	Tall, evergreen grass, forage supplement.
Alfalfa or Lucerne *Medicago sativa*	Perennial (5 years), N-fixing, forage.
Carpeting thymes *Thymus* spp.	Also called creeping thymes. Edible leaves, many varieties scented.
Lawn chamomile *Chamaemelum nobile*	Tough, hard-wearing ground cover. Grows in full sun or partial shade.

* spp. — species. It represents the plural form, i.e. several species. ** N-fixing — nitrogen-fixing.

Table 4.7. Examples of living mulch.

Green manure crops

Green manure crops are cover crops. This means that they are grown and then turned into the soil, or cut and allowed to lie on the ground surface. The whole idea is to allow plants to reduce weed growth, protect the soil and provide organic matter for the soil when the main crop is sown.

A catch crop is different again. These are fast-growing plants that are often interplanted between rows of the main crops or between successive seasons of crop (after main crop is harvested and before the next is sown). The classic example is radishes, which are planted between rows of other vegetables. The radishes grow so fast that they can be harvested well before the other vegetables mature.

Nitrogen-fixing

Nitrogen-fixing plants can convert nitrogen from the air into substances that are initially stored in the roots and then used throughout the plant. But the plants themselves cannot do this — they need help from certain bacteria.

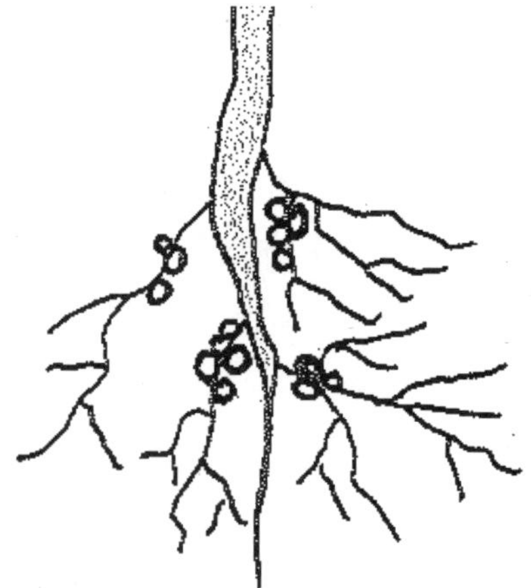

Nitrogen-fixing nodules on the roots of a shrub.

» DID YOU KNOW?

The term "pulse" is used to describe any nitrogen-fixing (legume) crops that are grown for their dry seed. This includes beans, peas, lentils and cow pea. Green peas and beans are also classified as vegetables. Some legume crops are also called grains.

There are many different bacterial species that fix nitrogen. These include cyanobacteria, which mainly live in water environments (coral reef); rhizobium, which inhabit legumes (peas, beans, alfalfa, clovers); and *Frankia*, which are present in alders, casuarinas and all species in the *Elaeagnaceae* or oleaster family.

Using nitrogen-fixing plants as the green manure crop increases the nitrogen concentrations in soils when these plants are killed or cut and dropped.

Seeds of warm season nitrogen-fixers. L to R: lentil, cow pea, mung bean.

Some of these plants suit warmer climates and others suit and tolerate colder climates, so these are grouped as warm season or cool season plants in Table 4.8.

» DID YOU KNOW?

Symbiosis occurs when at least two different species coexist and both benefit from the association.

It is very common in nature, and a well-known example is lichen, which is a combination of algae and fungi.

The algae is photosynthetic and makes the food for the fungus (which cannot make its own), while the fungus absorbs moisture to keep the algae alive (algae are seaweeds).

Warm season nitrogen-fixers	Notes
Cow pea *Vigna unguiculata*	High protein, annual food crop, hardy, tolerates shade and dry conditions. Grows in poor soil and drier conditions than many other warm season crops.
Lablab *Lablab purpureus*	Fast-growing annual, edible leaves, beans need cooking.
Lentil *Lens culinaris*	Edible pulse (seeds used as food), high protein content.
Mung bean *Vigna radiata*	Bean eaten as sprouts — raw or cooked.
Soybean or soya bean *Glycine max*	High-protein food, source of soy milk, tofu and oil.

Cool season nitrogen-fixers	Notes
Broad bean *Vicia faba*	Tall, annual plants, forage for animals, seeds cooked before eating.
Chick pea *Cicer arietinum*	High-protein seeds, makes hummus, nutritious.
Lupin *Lupinus* spp.	Tall, annual or short-lived perennial plant, forage, nutritious seed.
Pea *Pisum sativum*	Small annual plants, some varieties climbing.
Subclover *Trifolium subteraneum*	Underground seeds, self-regenerating annual, forage.

Table 4.8. Common nitrogen-fixing green manure crops.

Soybean

Grain crops

Grain crops are plants that produce small, hard, dry seeds that are harvested for animal or human consumption. They can be cereals (like wheat and barley) or legumes (like soybeans and peas). Cereals belong to the grass family; plants such as amaranth, quinoa and buckwheat are called pseudocereals because they are not true cereals but their grain is used as a staple food.

» DID YOU KNOW?

Plants have different photosynthetic pathways. Warm and cool season plants differ in their metabolic pathways and the way in which they use carbon dioxide during the photosynthesis process. They either utilize a three carbon molecule (a C3 plant) or a four carbon molecule (a C4 plant). A few agaves and cacti can utilize either mechanism (and are called CAM plants). Most plants are C3 plants and are adapted to cool season establishment and growth in either wet or dry environments. C4 plants are more suited to warm or hot seasonal conditions under moist or dry environments. C3 plants generally have better feed quality and tolerate frost, but C4 plants tend to demonstrate prolific growth. The succulents of the CAM group can survive in dry and desert areas.

Broad bean

Buckwheat

Warm season grain	Notes
Buckwheat *Fagopyrum esculentum*	Short season crop, tolerates acidic soils of low fertility but not flooding. Used for erosion control. Buckwheat is a pseudocereal and is not related to wheat or any other grasses.
Pearl millet *Pennisetum glaucum*	Very productive, small-seeded grass. Short growing season, common in tropical countries. Another type that is not related, Japanese millet, *Echinochloa esculenta*, is grown in poorer soils and cooler climates.
Sorghum *Sorghum bicolor*	Used for alcoholic beverages, biofuel, molasses, food for humans and animals. Drought and heat tolerant.
Sunn hemp *Crotalaria juncea*	Tropical N-fixing plant, useful fodder and fiber.
Sunflowers *Helianthus annuus*	Nutritious seed, good oil. Flowers attract birds and beneficial insects. Some varieties can be very tall. Productive plant.

Table 4.9. Examples of warm season grain green manure crops.

Cool season grain	Notes
Barley *Hordeum vulgare*	Common cultivated grain, used to make beer, malt.
Oats *Avena sativa*	Used as oatmeal and rolled oats by humans. Mainly used as livestock feed — both as green pasture and cut as hay. Oat straw for animal bedding.
Cereal rye *Secale cereale*	Rye grain increasingly used for bread, whisky, beer and animal fodder. Also called rye corn. Tolerates poor soils, cold climates.
Wheat, triticale *Triticum* spp.	High protein seed, used to make bread, able to be stored. Grows in low rainfall areas. Hay and straw used in buildings and for animal fodder. Triticale is a hybrid of wheat and rye and is more disease-resistant.
Canola *Brassica* spp.	Several cultivars mainly grown for oil from seeds. Related to broccoli, cabbage and turnip.

Table 4.10. Examples of cool season grain green manure crops.

Fermented fertilizers

When we grow food, we often neglect plant nutrition. We forget that plants take up nutrients from the soil and the soil becomes depleted.

In all agricultural systems, the addition of fertilizer to our soils is crucial to food crop production. We can make simple organic-based fertilizers from weeds, manures, worm castings and various combinations of any or all of these.

In these preparations, bacteria and other microorganisms are used to break down organic matter into simpler substances that can be applied to plants and the soil.

The digestion of plant and animal material can be undertaken without air (anaerobic) or with air (aerobic).

Fermentation, by definition, is an anaerobic process, but we use it here to include any digestion of organic matter, with or without air.

Anaerobic digestion

Some bacteria do not require oxygen to survive, and they are able to use plant and animal material as their food source.

When this occurs, various gases are produced as a by-product of the digestion process. These mainly include methane, carbon dioxide and nitrogen, but may also include sulfur oxides, hydrogen sulfide and ammonia.

All of these gases need to be vented from the system, otherwise gas pressure builds up and liquid and gas explosions can occur.

Manure is placed in a drum or container, water is added, along with some other ingredients such as molasses and milk, and allowed to ferment for at least one month and up to a few months depending on climatic conditions. Excess gases are passed out of the tank through a water trap, which excludes air from entering but allows metabolic gases to escape. If these gases cannot escape they build up in the system and inhibit microbe action.

» DID YOU KNOW?

Biofertilizer is a fertilizer that contains living organisms. You basically produce a microbe culture to inoculate the soil and a solution containing readily available minerals to feed the plants. You can "seed" a culture with a particular type of microorganism (e.g. rhizobium, blue-green algae), or you can make a generic brew that contains microbes that are found in the manure, yeast, air and water that you add. You need to use the fertilizer well before the culture dies. Even so, adding minerals and nutrients to plants is beneficial every time.

Biofertilizer is usually made from manure. Most animal manures seem to work, including chicken manure or other bird manures with mixed success. Manure from grass-eating animals, such as cows, sheep and goats, seems better than that from meat-eaters, such as dogs. Pig manure is used to produce methane.

When methane is the main by-product, it can be collected and used as biogas, a fuel that can be used for cooking and heating.

Anaerobic digestion — makes biofertilizer

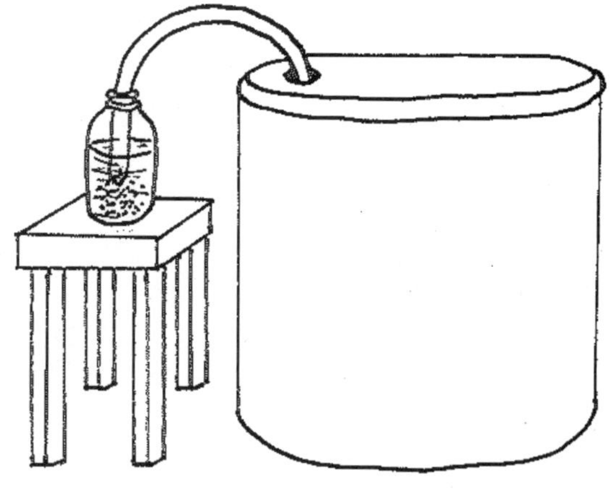

To make a brew, add about 45 lb of fresh manure to a 55 gal drum (with a lid that can be sealed). Fresh manure contains more beneficial bacteria and other microorganisms. Fill to about four-fifths with water.

Add small amounts of a range of other substances, including milk, yeast, rock dust and molasses, all of which help to kick-start and feed the microbe population. Install a water trap and allow to ferment for a month or two (visible bubbling may stop after a few weeks or more). When you open the lid, beware! — it may stink.

If it does smell badly then the product should not be used — it most likely contains pathogens. While biofertilizer does smell, it is not unpleasant — it should smell like a typical ferment.

Pass the liquid through a mesh screen or coarse cloth to remove the digested remains of the manure.

The resulting liquid fertilizer can be diluted at least ten times with water before spraying on the garden. You should be able to find some laboratories that can test your biofertilizer product if you are concerned about its contents.

You will also be able to find other recipes that make biofertilizer, some being specific for particular applications.

Aerobic digestion — makes compost tea

Aerobic digestion

Aerobic microbes need oxygen. Air is pumped into the solution by using an aquarium pump for smaller containers to air compressors (2–3 ft^3/min) for larger tanks. Intermediate bulk containers (IBCs) are cheap to obtain second-hand and are ideal. They hold about 275 gal.

The general term for the products from this process is "compost tea". You can suspend a bagful of earthworm castings in a 5 gal bucket and aerate for a day or two (generally no more than two days).

On a larger scale, a cereal bag (wheelbarrow-full) of animal manure or shredded weeds and plant material can be aerated by a larger blower or compressor. The drum is open to allow air to escape. Plant material should be finely shredded as much as possible whereas animal manure easily falls apart in water.

Compost teas are quick to make, but they must be used at once to maximize their effectiveness and there is an energy cost for the aeration.

Both compost teas and biofertilizer help build soil fertility, and these products can be made from most organic wastes.

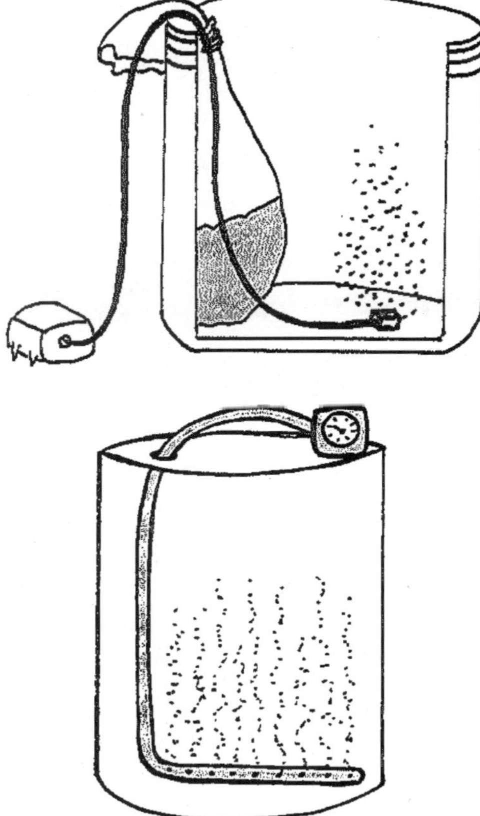

A large air compressor can aerate a larger volume of fertilizer.

Weeds as indicators of soil conditions

The promotion of weeds and plants to indicate certain soil conditions, or as dynamic accumulators that mine particular minerals, should be viewed with caution.

Much of the literature on the role of plants to absorb and store particular nutrients is more folklore than science.

Popular websites and books list tables of dynamic accumulators and lists of common weeds and what they suggest about the structure of the soil, without any reference to sources of information.

There is some evidence, of course, but there appear to be occasional contradictions. Do the plants take up particular nutrients even when these are deficient in the soil, or do the plants simply take up these nutrients because they are abundant in the soil?

Some people suggest that the mineral makeup of plants is a reflection of what's in the soil, so a plant might have high boron levels because this is present in that soil.

» DID YOU KNOW?

Human body waste is also broken down by anaerobic and aerobic bacteria.

Toilet and other household effluent are decomposed without air in a septic tank system, while on-site aerated treatment plants or municipal wastewater treatment plants use a combination of anaerobic and aerobic bacteria to break down our wastes before discharge or use.

In many cases the treated effluent is used for irrigation.

The remaining sludge and solids are often dried and further composted and changed into various amendments and additives for soil amelioration.

Much of the information about accumulators and indicators is based on historical observations and anecdotal evidence.

Sometimes people do compare chemical soil analysis results with visual observations about the distribution and abundance of weeds in that soil, and I have done this many times, but there is very little research to date about these things.

For example, chickweed is reputed to be an accumulator of manganese and copper, but there doesn't appear to be any scientific research to support this.

On the other hand, I have observed capeweed in areas of known calcium deficiency, dock, sorrel and plantain in acid soils, barley grass in saline soils or where the water table is high, stinkwort in dry, clayey soils, and lamb's quarters in rich cultivated soils in summer.

That is not to say that these types of plants are only found in these conditions, so much more research has to be undertaken before we can be confident about using weeds, herbs and other plants as indicators of soil health.

Simple soil tests

Monitoring and measuring soil components helps us know what amendments to add and what fertilizer we may need to apply. The backyard gardener can perform regular checks on simple parameters such as acidity, and measure the composition of the soil as they endeavor to change it.

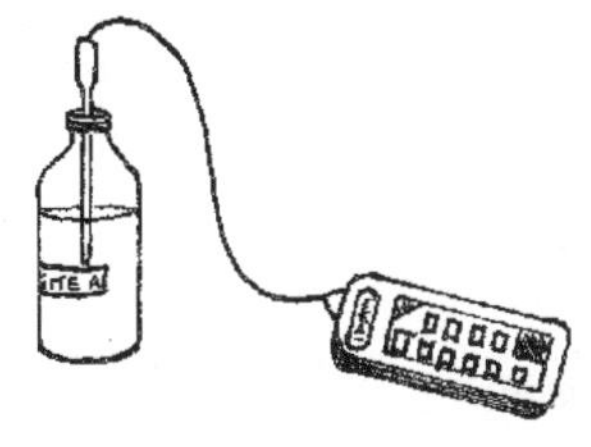

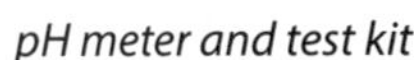

pH meter and test kit

Acidity is measured with a pH test kit. pH is the numerical concentration of hydrogen ions placed on a scale 0 to 14, with 7 meaning neutral (not acidic or alkaline = pure water); numbers less than 7 are acid and those greater than 7 are increasing levels of alkalinity (basic).

These numbers are logarithmic, so there is a factor of 10 from one number to the next. For example, pH = 5 is 10 times more acidic than pH = 6.

Most plants like a pH somewhere between 6 and 7.5, although blueberries prefer 5 to 5.5, while asparagus and calendula tolerate 7.5 to 8.5.

An electronic pH meter or a powder test kit can be purchased and easy instructions will enable you to measure the acidity of a soil (or a water source too if a pH meter is used).

Soil composition can be determined by using specific-sized screens (sieves) or a volumetric cylinder. A sample of soil is passed through a series of screens to separate stone, sand, silt and clay. The amount of each left on each screen will give you a good indication whether the soil is a clayey-loam or a sandy-loam, and so on.

If you have a 500ml or 1l laboratory glass cylinder, you can add soil to the ½ to ¾ way marks. Add water to nearly the top. Stopper and shake to mix thoroughly.

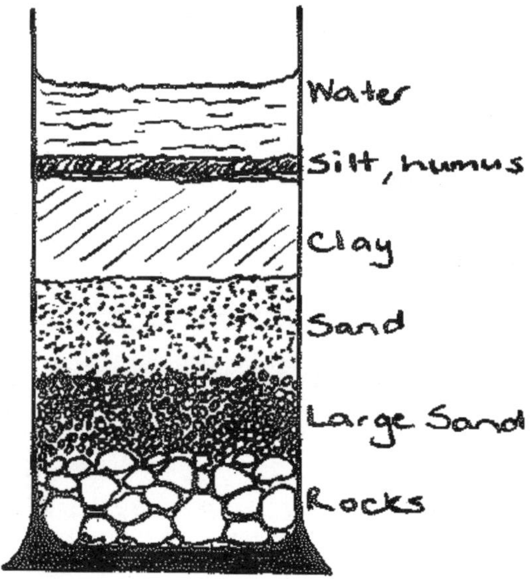

Soil profile. Shake soil with water and let settle.

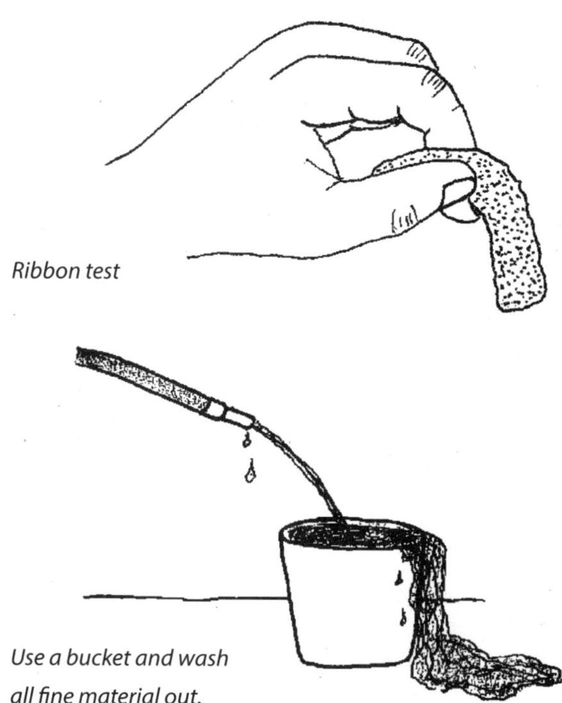

Ribbon test

Use a bucket and wash all fine material out.

Allow the particles to settle and you can generally see layers of floating organic matter, settled organic matter, fine clays and silts and then coarser sands and stones at the bottom.

The ribbon test is used to determine soil composition too, and provides information about the clay content.

Here, a small sample of soil and water is mixed in our hand, molding and kneading until a 1 in ball is formed. This is then pushed out between the thumb and forefinger (first finger) into a thin strip.

Measure (estimate) the length of the strip or ribbon when it falls off. Soils with reasonable clay content will produce a 2–3 in ribbon.

A chart that allows you to interpret the soil composition is readily available from books and the web. The feel of the soil while you are molding it also gives us clues about its texture and composition.

Clay content can also be determined by the bucket method. Place 2 lb (2 l) soil in a laundry 2.5 gal bucket.

Wash the soil with a garden hose and continually stir with one of your hands. The finer clay and silt particles will wash over the edge of the bucket, leaving sand and stone behind.

Once the water is relatively clear, we can assume that clay and silt have been washed out.

Remeasure (volume or weight) what is left behind and you can do some math to determine the percentage clay content of the soil. This is the test to perform if you want to make mud-bricks (50–80% clay required) or a rammed earth wall (20–50% clay).

Other soil tests

While you can do simple tests yourself to determine the acidity and the relative amounts of different particles in a soil sample, you can also send soils to laboratories where they can determine macro- and micronutrient levels, the cation exchange capacity and the degree of salinity.

This type of analysis is important if you want to ensure the correct amount of amendments are added to your soils.

» DID YOU KNOW?

The Cation Exchange Capacity (CEC) is a measure of the soil's ability to hold cations. Cations are positively charged atoms and include three important plant nutrients: calcium (Ca^{2+}), magnesium (Mg^{2+}) and potassium (K^{+}).

Cations are held by negatively charged particles, called colloids, which are common in clay and humus. The humus found in organic-rich soils can hold about three times more cations than the best type of clay.

As plants take up the cation nutrients, other cations take their place on the colloid. The stronger the colloid's negative charge, the greater its capacity to hold onto and exchange cations (hence the term Cation Exchange Capacity).

The CEC is a good indicator of soil fertility, as a high CEC suggests many nutrients may be available to plants.

EDIBLE FOOD PLANTS 5

THIS SECTION COULD HAVE BEEN VERY LARGE, as we eat literally hundreds of different types of plants throughout the world. However, it is confined to a select few of the most nutritious, easy-to-grow plants that we so often eat as food: vegetables and garden salad fruits.

Top ten vegetables

Vegetables are typically the roots, stems, tubers, corms, leaves and flowers of a plant that we eat. Anything that contains seeds we will call fruits. So, peas, beans, peppers and tomatoes, which are often labeled as vegetables, are, in fact, fruits.

Not all vegetables can be grown in all climates, but there could be cultivars that are best suited to your region. I am sure there are some vegetables listed here that you may never have tasted, or your children or other family members might not like to experiment (to taste and see) as much as you do.

The vegetables are listed alphabetically rather than by nutrition levels or any other ranking. There is debate to the benefits of each of these and it is easy to form a list of other vegetables that could claim a spot in the top ten. The criteria for this list are that they need to be:

- Common and affordable — readily available, reasonably cheap and can be bought at local markets or grown in most countries.
- Nutrient-dense — need to contain a fair amount of most of the vitamins and minerals required for good health. The level of nutrients is high compared to the number of calories the food contains.
- Whole foods and do not contain any additives — should taste good, do not need any processing and are easy to prepare and cook.

Globe artichoke

Asparagus spears

Broccoli

Artichokes (*Cynara cardunculus*)

There are two types of artichokes, which have no botanical relationship at all. The artichoke discussed here is the globe artichoke, which is a large thistle, and the flower or bud is eaten. The other plant, the Jerusalem artichoke (*Helianthus tuberosus*), which has nothing to do with Jerusalem or artichoke really, is a member of the sunflower family. It too is simple to prepare and use, nutritious with good levels of some vitamins and minerals (high iron), and has little or no starch, unlike potatoes, but is high in fructose (which is much better for diabetics than sucrose).

The globe artichoke is high in vitamins C and K, folate, B6, magnesium, potassium, calcium and manganese. The bud is picked before the flower starts to open, the outer sepals peeled away and the inner part boiled for about 20 minutes.

Asparagus (*Asparagus officinalis*)

Fresh, young asparagus shoots are hard to beat for taste. No need to cook them, just eat them raw. Asparagus contains vitamins C and K, folate, B1 and B2, and tryptophan (tryptophan is an essential amino acid crucial for protein (enzyme) production). Asparagus does not contain any fat or sodium.

Broccoli (*Brassica oleracea italica*)

Broccoli is good steamed, raw (as a snack or in salads) and cooked in soups, stir-fries and other dishes. Both the green flower heads and stalks are nutritious.

Broccoli contains high quantities of vitamin C (twice that of oranges), soluble fiber, a fantastic supply of most other vitamins and minerals (about half the calcium found in milk) and a number of other substances that help us fight disease.

Carrots (*Daucus carota*)

Carrots are high in fiber, vitamins C, K and B6, and minerals potassium and manganese. Well known as an exceptional source of vitamin A (one large carrot provides your daily requirement). They are slightly sweet to taste due to the presence of sugar, but are also high in carotenoids such as beta-carotene and a wide range of other antioxidant phytonutrients. Grate them on top of a salad, and put small (baby) carrots into your lunchbox for a between-meal snack. Carrots prefer to grow in sandy soil, and if grown in clayey or heavily composted soils they tend to stunt.

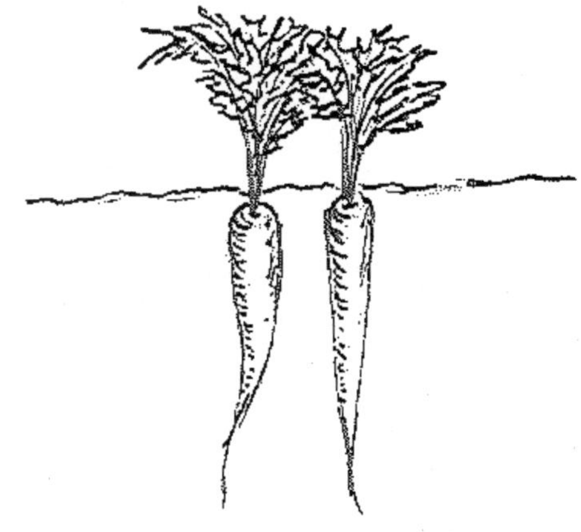

Carrots

» DID YOU KNOW?

There are lots of different carotenoids in organisms and their roles range from pigments that absorb light for photosynthesis, to protecting organs from damage, to antioxidants that protect us from substances that cause cell damage.

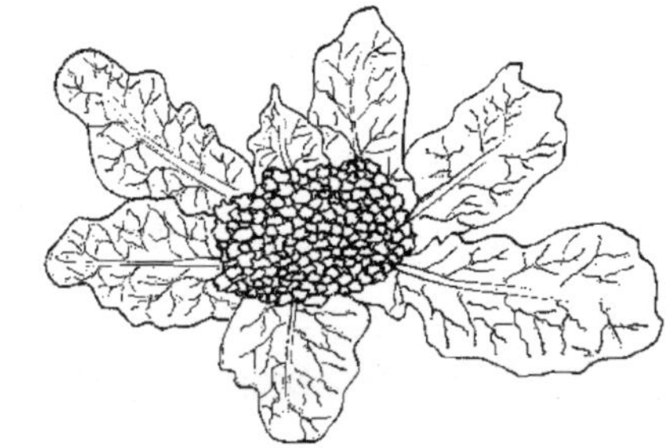

Cauliflower

Cauliflower (*Brassica oleracea botrytis*)

As the name implies, what you eat of the cauliflower is the immature flower stalks and buds. Cauliflower is high in fiber and vitamin C, with good amounts of vitamins K and B6, and minerals potassium and manganese. It contains no fats but does have fair levels of omega-3 and omega-6 fatty acids.

Cauliflowers can be a little bland to eat, but can be eaten raw and are a great addition to a plate of dips and relishes.

Garlic (*Allium sativum*)

Garlic has a traditional use as a medicinal plant, and it does contain high levels of vitamins C, thiamine (B1) and B6, and the minerals manganese, calcium, selenium and phosphorus. While some people eat a clove of garlic each day, it can be added to most stir-fry dishes, soups and stews.

Garlic

Kale

Kale (*Brassica oleracea acephala*)

Kale is in the same family as broccoli, cabbage and Brussels sprouts, but not as well known. It has exceptional levels of vitamins A, C and K and very good amounts of vitamin B6, manganese, copper, potassium, calcium and iron. Like cabbage, the leaves are cooked, and kale chips can be made by sprinkling oil and salt over the leaves and then drying them in a low-temperature oven.

All of the brassicas are nutritious, and as a group probably the most nutritious of all vegetable groups. A special mention should be made of Brussels sprouts (*Brassica oleracea gemmifera*). This vegetable, probably hated by most children, is a rich source of vitamins A, C and K, folate, manganese, fiber and carotenoids. Brussels sprouts are best quickly steamed to preserve nutritional values and avoid releasing sulfur smells caused by excess cooking.

» DID YOU KNOW?

We need various minerals in our diet to function properly. Iron is found in every red blood cell, calcium and phosphorus in all of our bones and teeth, magnesium helps muscles contract and relax, potassium and sodium make nerves send messages, and zinc make enzymes work better.

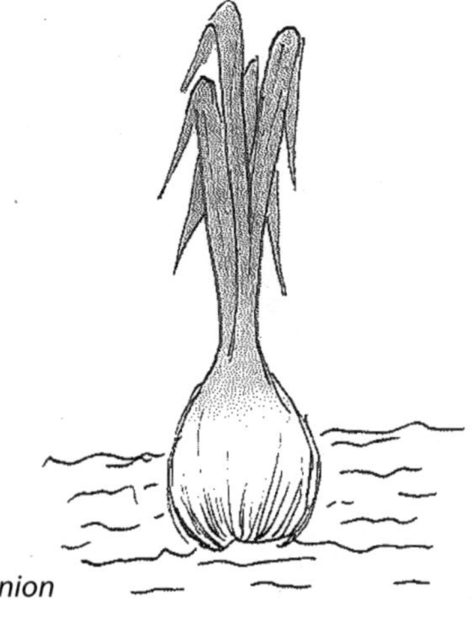
Onion

Onions (*Allium cepa*)

Can be eaten raw in salads or as a flavoring for many cooked dishes. While they taste sour, they do contain sugars. There are reasonable amounts of most vitamins and minerals, and onions are high in fiber. Onions take a long time to grow and mature, but are generally disease free and hardy.

Spinach (*Spinacia oleracea*)

This vegetable should not be confused with silverbeet and chard (*Beta vulgaris*), and the spinach here is English spinach, although it originated from central and western Asia.

Historically (and promoted through the cartoon Popeye) spinach was considered to be one of the best sources of iron, but it only supplies about 5% of the daily requirement. It does, however, contain excellent amounts of vitamins K and A, as well as good amounts of folate and vitamin C, and minerals manganese and magnesium.

Spinach is best eaten raw in salads, but can be lightly steamed or added to soups.

Spinach

Sweet potato (*Ipomoea batatas*)

Exceptional amount of vitamin A, good amounts of potassium, copper and manganese, and a fair amount of other vitamins and minerals.

Sweet potatoes can substitute for the common potato and can be used as chips, mashed with vegetables, baked for roasts and even added to soups.

Sweet potatoes can be grown under glass or in polytunnels in cooler climates.

The main types are the orange- and white-fleshed varieties (they do taste and cook differently), and unlike the common potato, the leaves are also edible and are a useful addition to stir-fries and cooked dishes.

Sweet potato — edible tubers

» DID YOU KNOW?

The sweet potato is not related to the common potato, *Solanum tuberosum*, which is closely related to tomatoes and peppers. Both plants have edible tubers, but while the potato is a swollen underground stem, the sweet potato is a swollen root. While the common potato is nutritious too, it doesn't have the same vitamin and mineral content. Potatoes are higher in starch and lower in sugars than sweet potatoes.

Cultivation notes

You can read all about how to grow plants from other sources, so here are just a few simple handy hints for each vegetable.

Vegetable	Cultivation
Artichoke	Summer crop. Pick the flower buds before they start to open. Leave a few buds to develop so you can save seeds, and you will be pleasantly surprised with the blue-purple flowers.
Asparagus	Summer crop. Cut the spears when they are about 4–6 in tall. Leave them too long and they go woody.
Broccoli	Winter crop. Once the main head is removed, continue to allow plant to grow, as many smaller secondary heads appear which are just as nutritious.
Carrots	Summer crop. Carrots are best grown in well-draining sandy soils, and will often stunt and distort when grown in richer organic soils.
Cauliflower	Winter crop. Like broccoli, cauliflowers produce smaller secondary heads that can also be picked and eaten.
Garlic	Winter crop. Usually harvested in spring and early summer when the stalk either starts to produce a flower head or dies off.
Kale	Winter crop. This is a picking green — pick a few leaves at a time for meals. New leaves grow and thus provide a long harvest time. Grow the purple-leaved variety as well as the green-leaved.
Onions	Winter crop. These take a long time to develop and are usually harvested in summer. Bend the stalks over (just use your foot) early in summer and allow onions to finish swelling as the stalks die off. Seeds don't keep well, so you need to sow them within six months of harvest.
Spinach	Winter crop. Also a picking green — pick leaves as you need. When the plant "seeds" it is time to remove from the garden bed.
Sweet potato	Summer crop. A great ground cover. Keeping the ground covered is an important permaculture idea. We can spread various materials over the soil, but there are also plants that will make a living ground cover. Sweet potato is easy to grow in warm conditions. Buy a tuber and place it in a container with water half way covering it. Soon shoots and roots appear. When the shoots are about 4 in long, gently detach them from the tuber and plant them. If there are adult sweet potato vines around you can take cuttings from them. Anywhere along the vine where there are new leaves growing, make a cutting about 4 in long and put it in a glass of water for a few days. New roots form quickly. Sometimes it is difficult to find the tubers, so when first sown, put in a garden stake to mark the expected tuber area. You can also simply lift and follow the vines to where they enter the ground and start digging.

Table 5.1. Cultivation notes for some nutritious vegetables.

Top ten garden salad fruits

When people think of fruit, apples, pears, bananas and mangoes spring to mind. There are many other fruits, often confused and grouped with vegetables, that I call garden salad fruits, so named because these plants are often eaten raw, in salads or cooked and prepared like vegetables.

These fruits include tomatoes, peppers, cucumbers, peas and beans, and as mentioned before, all of these culinary delights contain seeds and they are chosen not so much for being the most nutritious foods ever, but because they are easily grown or obtainable (besides being very nutritious of course).

Here are my top ten garden salad fruits, listed alphabetically.

Beans (*Fabaceae* family)

The most common bean found in the supermarket is the green bean (*Phaseolus vulgaris*), also known as the string bean and snap bean. These are not as nutritious as other, lesser-known beans, which are discussed here. Green beans only have about 10% of the nutritional value of lima and broad beans.

Green beans

Lima beans (*Phaseolus lunatus*, also known as butter beans) are a high protein, high fiber, low fat food. They have exceptional amounts of thiamine, folate, magnesium, potassium and manganese, and a good range of other vitamins and minerals. Raw lima beans, like many other legumes, contain cyanide so these must be cooked before eating.

Broad beans or fava beans (*Vicia faba*) also have high protein, fiber, folate and manganese, with very good levels of magnesium, copper and phosphorus. Not everyone likes their taste, and sometimes these beans are just grown for animal fodder or as a green manure crop and turned into the soil.

Cantaloupe

Cantaloupe (*Cucumis melo*)

Known as rockmelons in Australia and muskmelons in the Middle East and Asia, the cantaloupe is a very nutritious food. It contains very high levels of vitamins A and C, and good levels of most other vitamins and minerals, including vitamin E, riboflavin, potassium and copper.

Unfortunately, most of the carbohydrates it contains are sugars. Cantaloupes have a ribbed skin, but a popular smooth-skinned variety is the honeydew.

» DID YOU KNOW?

Watermelons (*Citrullus lanatus*) are very nutritious too. They also contain vitamins A and C and many other nutrients, but in smaller amounts than those found in cantaloupes. About 80% of the carbohydrates are sugars, hence the sweet taste.

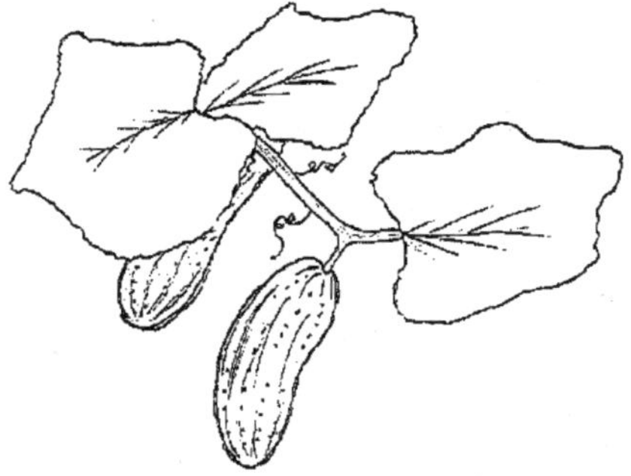

Cucumber

Cucumber (*Cucumis sativus*)

Related to rock melons and watermelons, cucumbers provide good levels of vitamins K and C and minerals molybdenum, copper and manganese. They are very low in saturated fat and cholesterol, and low in carbohydrates (starch and sugars).

Smaller, thin-skinned cucumbers (gherkin variety) are used as pickles, while the thicker-skinned, larger varieties are usually peeled and eaten raw in salads. A third variety is the burpless, which are thinner skinned and a little sweeter to eat. Some varieties are seedless.

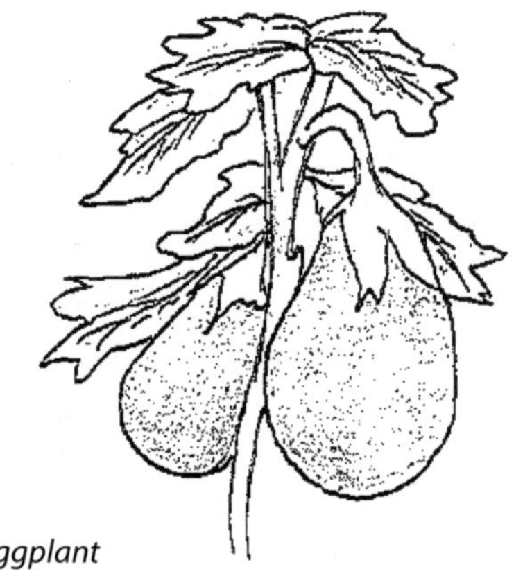

Eggplant

Eggplant (*Solanum melongena*)

Also known as aubergine, eggplant contains reasonable levels of fiber, vitamin B1 and manganese, and fair amounts of most other vitamins and minerals. Eggplants also contain high levels of phytonutrients, which have antioxidant properties to help fight disease.

Peas (*Pisium sativum*)

Peas are climbing annual plants. Green peas are the most common variety and these are best eaten raw or lightly steamed. Snow peas (*P. sativum* var. *saccharatum*) and snap peas (*P. sativum* var. *macrocarpon*) are both picked when the pods are still tender and these, along with the seeds, are eaten raw and chopped up and cooked. Fresh peas can be added to salads, soups and stir-fries.

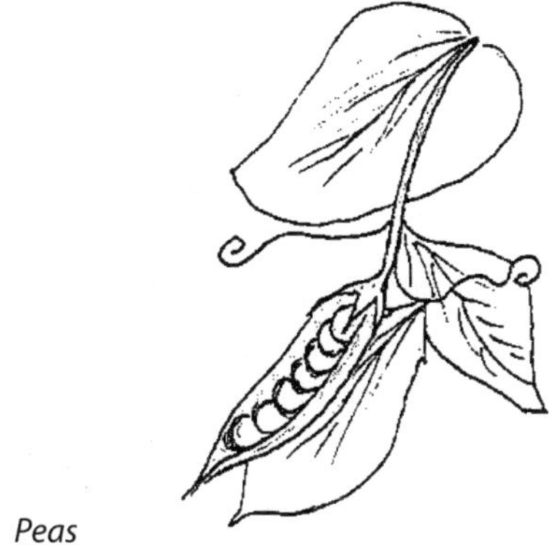
Peas

Peas are good sources of vitamins C and K, thiamine, folate, manganese, zinc, iron, magnesium and phosphorus. Mice can be pests and will eat young pods.

Peppers (*Capsicum annuum*)

These are large mild forms of the pepper group. Spicy (hot) varieties are the smaller chili peppers while the larger red and green varieties are known as bell peppers in the Americas and capsicum in Australasia. Other countries just refer to them as peppers. The green variety is the fruit in its immature state, and it will often change into the red, yellow or orange ripe state if left on the shrub. The red form is more nutrient-dense than the green form, but both of these are generally more nutritious than the hot peppers.

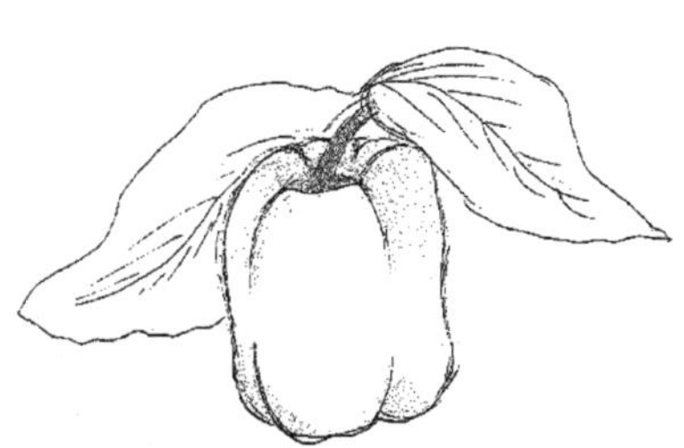
Bell pepper

They are very high in vitamin A and C, reasonably high in vitamins E, K and B6 and minerals potassium and manganese. Two-thirds of the carbohydrates are found as sugars, so bell pepper is also known as sweet pepper.

Peppers are great in salads and stir-fries, and can be roasted or cooked in hot dishes. They are usually grown under glass in far northern or southern areas.

» DID YOU KNOW?

Whenever you grow vegetables let one or two plants "go to seed." Saving your own seed enables you to grow those varieties you love to eat.

When storing seed, don't keep them in a paper bag or a jar for too long. After a few years you can assume most seeds are no longer viable. It is always best to plant seeds within a year or two to ensure good germination rates.

This is true of most vegetable seeds. For example, onion and garlic seeds last about one year, and peas and beans maybe up to three years. Donate your excess seed to a local seed bank or a seed exchange network.

Pumpkin

Pumpkin (*Cucurbita pepo*)

An underrated food, as it contains a very high level of vitamin A, as well as fair levels of most other vitamins and minerals, and is low in fat and sugars. There are so many varieties that you should be able to find a few that can grow in your climate and soil. Most varieties store well and thus can be used throughout the year.

Pumpkins, squashes, gourds and zucchini (courgette) all belong to the same genus, and many are cultivars of essentially the same species.

Pumpkins have both male and female flowers, with males appearing a few weeks before female flowers. Lots of flowers doesn't necessarily mean lots of pumpkins, as female flowers have to be pollinated.

Squash (*Cucurbita maxima, C. moschata* and *C. mixta*)

The terms squash and pumpkin are often interchangeable. This is because they are so closely related. What Australians call butternut pumpkin, North Americans know as butternut squash.

Squash

Squashes can be grouped as either winter or summer squashes. Winter squashes are thick skinned and store well, so they are grown in summer for eating during winter, while summer squashes are normally harvested while still immature (thus smaller) and eaten soon after.

Winter squashes have good levels of fiber, vitamins A and C, and minerals manganese and potassium, and tend to have higher levels of vitamins and minerals than summer squash varieties.

Tomatoes (*Solanum lycopersicum*)

One of the most common foods eaten in salads during summer, made into sauce and cooked in soups and other meals, tomatoes contain good levels of vitamins A, C and K, and minerals potassium and manganese, while being very low in saturated fat and cholesterol.

Sun-dried tomatoes are even more nutrient-dense and possess concentrated amounts of protein, fiber, and most of the vitamins and minerals (including sodium, which is not so desirable).

There are many varieties and cultivars that tolerate all types of climates. Even though most people associate red with tomatoes, cultivars of tomatoes are now found in a range of colors. Some stay green but are ripe, others are orange, yellow, purple-brown and some are striped and marbled. The shapes change too, and you can find very small round ones to oblong or squarish-looking ones.

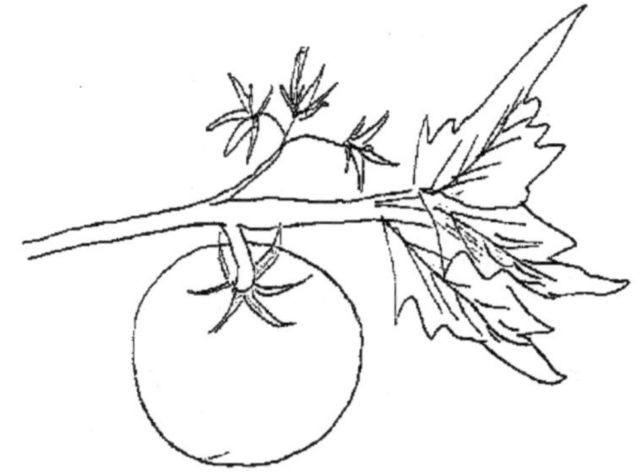

Tomato varieties can be different shapes.

» **DID YOU KNOW?**

The tomato, eggplant and potato belong to the nightshade family. Many of these plants are poisonous, especially when the fruits are eaten "green" and unripe (raw). They contain alkaloid toxins such as tomatine and solanine.

Leaves are always poisonous. Frying and cooking the green fruits does reduce the toxins they contain.

If the tubers of potatoes start to turn green you should discard them as the levels of toxins are starting to build up (plant them in the garden for another crop next season).

Zucchini (*Cucurbia pepo* var. *cylindrica*)

The zucchini or courgette is a summer squash and is a cultivar of the pumpkin group. It contains reasonable amounts of vitamins C, B6, riboflavin and folate, and the minerals manganese and potassium. Zucchini flowers can be eaten — they are usually deep fried or stuffed with other vegetables and cooked. Like pumpkins, zucchini store well.

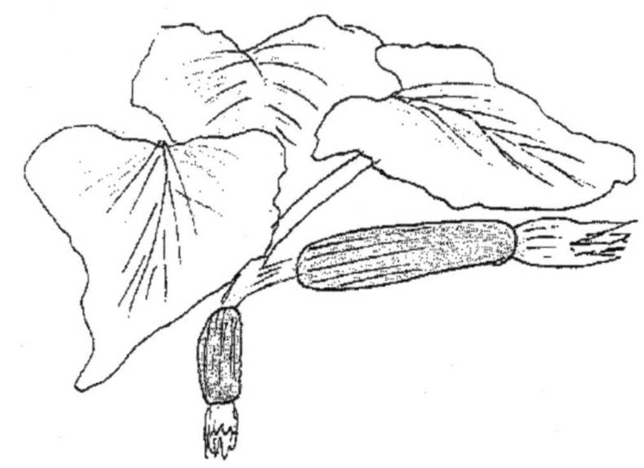

Zucchini

Cultivation notes

Every plant has specific requirements to grow. Again, some of these plants may not be suitable in your area, so find out what plants are more appropriate and research their nutritional value.

Garden salad fruit	Cultivation
Beans	Sow in spring for summer and autumn harvest. The productive life of the bean plants can be extended by continuing to feed liquid manure fertilizer regularly, even into the picking time. These fix nitrogen so leave their roots in the ground when you remove the plant. Let the last pods remain on the best bushes to dry for next year's seeds.
Cantaloupe	Likes a warm climate to grow and won't survive in frosty conditions. Doesn't store well even in a refrigerator, so you need to eat within a week after picking.
Cucumber	These vines can be grown on a trellis or mesh frame to keep the fruits off the ground. The cucumbers will hang down and will be easy to pick.
Eggplant	Grows as a perennial in tropical regions and more of an annual in temperate climates. Self-supporting shrub, but if the fruit get too big and heavy and touch the ground, stake that branch otherwise pests will eat into it.
Peas	Most pea varieties are fast-growing annuals and need staking or a frame to climb up and attach to, as they are tender vines. Most cultivars are picked when the pods are full and green, but others are left to dry out and they become split peas for soups.
Pepper	Small upright shrub that prefers full sun and a long summer to mature. Grows easily from seed. Heavy feeders when young, but taper off fertilizer use as they set fruit.
Pumpkin	Mainly annual plants with the vines requiring some room to spread. Common practice is to plant 4–5 seeds around a small mound, and thin these to the strongest 3 plants. Don't push seeds too deep into the soil as they won't germinate.
Squash	All cucurbits (pumpkins, squash and zucchini) have separate male and female flowers. Bees are required to transfer male pollen to the female flower. If pollinators are in short supply you may have to hand-pollinate to ensure that fruit form.
Tomatoes	Most tomato varieties need staking to hold up their branches and fruit off the ground. They love full sun and don't do well in areas that have frequent frost.
Zucchini	Very easy to grow. Tends to be a spreading shrub rather than a creeping vine like most squash. Sometimes you can get too many fruit on the one plant, so to reduce the possible over-abundance remove some of the flowers.

Table 5.2. Cultivation notes for garden salad fruits.

Handy hints

The following are some observations and handy hints to help you grow food plants successfully:

- Make sure you grow your food crops close to your house (zone 1 in permaculture). If you are constantly walking past your garden beds or looking at them through the kitchen window you will be more inclined to visit the garden, maintain the crops, pick the fruit and vegetables and ensure they get enough fertilizer, water and mulch.
- While we tend to grow particular vegetables at particular times of the year, don't be restricted by what seedlings are available or what generally is provided by horticultural people. Tomatoes are a typical summer crop but they can grow in winter too and even get a crop. Sure it takes a little longer to ripen and maybe the yield is not as great, but give it a go.
- Lettuces are usually grown in the winter and spring seasons and are picked early summer. If the temperature gets too hot they "bolt" and start to produce seed stalks and the leaves toughen up. But you can still grow lettuce in summer provided they are protected from full sun and you simply use the plant as a plucking green — removing leaves as you need for a salad.
- You really do need to experiment in your area and climate because local drought, flood, fire, hailstorms and temperature extremes will all affect what you grow and when.
- When books or people say "grow in full sun" this really means four or five hours of daily sunlight. Not many plants like baking in the summer sun, so most vegetables should be grown in dappled sunlight or partial shade or in direct sunlight for only a short time.
- While generic descriptions of some vegetables have been described in this chapter, search for local varieties that are known to grow well in your area.

  For example, a French variety of pea has small seeds that don't fatten up and the pods are flat and wide. The seeds (peas) are not harvested but the whole pod is picked and eaten, hence the name *mange-tout*, meaning "eat the lot."
- Endeavor to find varieties that might store well. Varieties of pumpkin, zucchini and squash often fall in this category, but tomatoes, eggplant and broccoli will store for longer when placed in a refrigerator.

- For cooler regions of the world consider building a dedicated polytunnel or glass hothouse for your winter vegetables. Then you should be able to grow cucumbers, tomatoes, Jerusalem artichokes and even cantaloupes out of season.
- Hygiene is crucial for successful food production. Remove fallen fruit, pick diseased leaves and prune when required to maintain your plants and minimize the threat of disease.
- Not all plants grow in the same soil. We know this but still get a "veggie" mix from the local nursery or landscaping business and place every sort of plant in it.

You will have better success when you grow your food plants in the right soil conditions. For example, carrots in sand, asparagus in rich, well-composted material, and garlic and onions in mildly acid soils.

FRUIT AND NUT TREES 6

ANY FRUIT IS A GOOD FRUIT. It was extremely difficult to choose the top ten fruit and nut trees, because there are so many in the marketplace, and so many are nutritious.

It is well documented that eating four or five pieces of fruit a day can reduce the risk of cancer, stroke and heart disease. Yet the upward nudge in fruit consumption is barely noticeable. Why? It surely must be about education and what parents are giving their children. How do we encourage parents to investigate and adopt a healthy eating lifestyle?

In keeping with the theme of grouping trees into their preferred growing areas, this section lists fruit and nut trees according to their optimum climatic zone. While some trees will adapt and grow in dryland climates as well as temperate climates, it is rare that tropical trees will grow elsewhere, unless particular cultivars are bred for this purpose.

The focus here is also on human foods, as fodder for animals is discussed later. As an example, carobs can be both human food and stock feed (in dry climates), but they are discussed in the chapter on fodder for farm animals as few humans eat the products from carob powder, although this is slowly changing.

I have included at least two nut trees for each climate zone, although some plants have both edible fruit and edible seeds. The lists were compiled from basic nutritional data, including levels of particular vitamins and minerals, but there has been no consideration of amounts of phytochemicals, carotenoids, flavonoids and polyphenols and a host of other substances commonly found in plants that are linked to health, antioxidant activities and prevention of disease.

For example, apples contain high levels of antioxidants but are much less nutritious than many other fruits. Blueberries have exceptional amounts of disease-fighting chemicals but just miss out in a spot in the top ten for the most all-round nutritious fruits.

» DID YOU KNOW?

The energy contained in fruit consists of sugars, and is digested in about 30 minutes.

The energy animal products contain consists of fat and proteins. This needs about six to eight hours to be digested. Fruit has high levels of fiber, animal products have no fiber, and the water percentage of fruit is about 80%, whereas that of meat is 15%.

Dryland fruit and nut trees

For simplicity, dryland climates are those with annual rainfalls of less than 24 in, often with hot summers and high evaporation, so we are looking for plants that survive in these harsh conditions.

The following plants are tough and survive well in low rainfall areas. Some of these plants will need supplementary watering, especially during fruiting and their growth stages, and during any long, dry periods.

Some of these plants are more often found in the cooler climate regions and make obvious choices for planting in this zone.

About half of the following fruit and nut trees are deciduous, and this may help to explain why they can grow in dry, desert areas, where nighttime temperatures can fall below freezing and trees with foliage could be severely damaged.

Almond (*Prunus dulcis*) [syn. *P. amygdalus*]

Almond

The almond is a deciduous tree that grows to 20 ft. Almonds are like woody peaches, with the flesh of the peach replaced by a leathery coat around the shell and seed. Almonds are incorrectly labeled as nuts, because the seed is large and eaten, but it is botanically a drupe — a plant that produces a fruit with a central seed inside. Stone fruits (e.g. plum, cherry, peach, apricot) are all drupes.

Almonds have a good range of the most common vitamins and minerals, and are high in vitamin E, riboflavin, protein, fiber, manganese, calcium and magnesium.

Apple cactus (*Cereus repandus*)

Tree-like, columnar cactus that produces red fruit with white flesh similar to dragon fruit. Also known as Peruvian apple. Under-researched nutritional value, but the sweet fruit has had a long traditional use.

» DID YOU KNOW?

Cacti do not have true leaves. The green stems carry out photosynthesis. The leaves have been reduced to become spines (see chapters 8 and 10). No doubt the thorns also protect the plant from browsers.

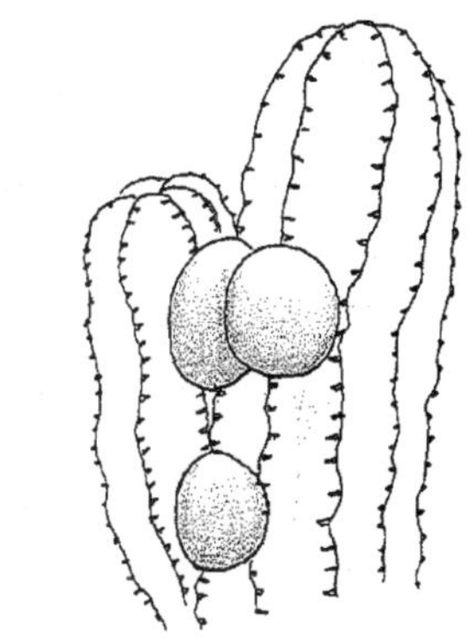

Apple cactus

Prickly pear (*Opuntia* spp.) is another possibility as they can be a great source of fiber, vitamin C and magnesium, and most species have mild amounts of other vitamins and minerals.

However, they are invasive and have been declared as noxious weeds in many regions and states.

Dragon fruit (*Hylocereus* spp.)

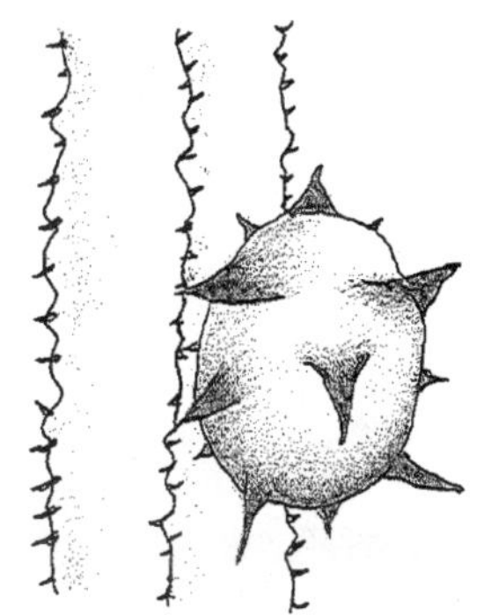

Dragon fruit

Creeping cacti, also known as pitahaya, with red, white or yellow flesh depending on the species. The white-fleshed, red-skinned variety, *Hylocereus undatus*, is the most popular. High in fiber, vitamin C and calcium, with fair amounts of B vitamins, phosphorus and iron.

Fig (*Ficus carica*)

A large-spreading, multi-stemmed shrub, fig trees are deciduous and produce false fruit. The "fruit" is not the traditional ripe ovary and seeds, but a bunch of flowers and multiple ovaries on the inside on a hollow fleshy stem.

Figs have small amounts of most vitamins and minerals, including potassium and manganese, and vitamins B and K. They are great to eat fresh off the tree, but equally taste fantastic when dried or made into jam.

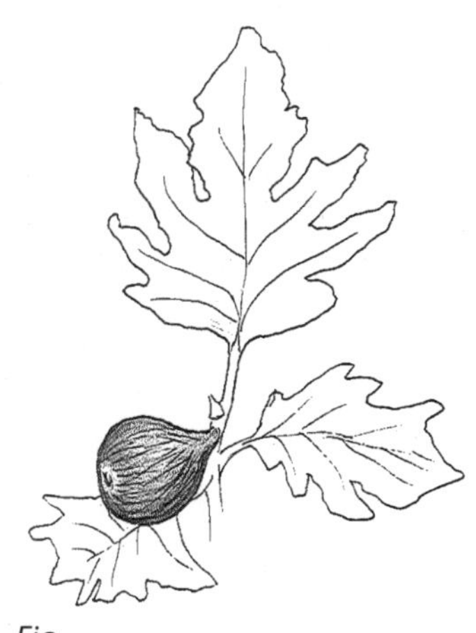

Fig

Jujube (*Ziziphus jujuba*)

Jujube is also known as Chinese date. The most remarkable thing about this fruit is that the flavor changes as it develops. You can eat when young, firm and green and it tastes like an apple. As it turns brown and shrivels up it tastes more like the common date.

While not having exceptional amounts of nutrients, it does have high vitamin C, and contains no fat and cholesterol. As it dries the

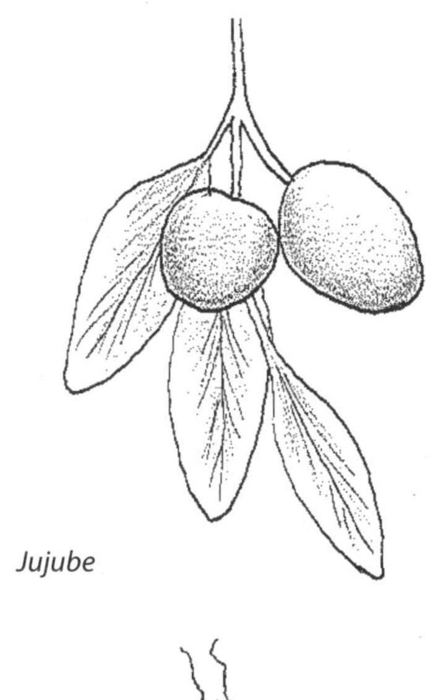
Jujube

relative concentrations of nutrients change, with vitamins generally decreasing and mineral content increasing. So dried fruit has proportionally more potassium, calcium and magnesium.

It survives in low rainfall areas, without additional watering once it becomes established. Unfortunately, this deciduous tree has traditionally been grafted onto thorny rootstocks that sucker. Non-suckering varieties are available.

Kei apple (*Dovyalis caffra*)

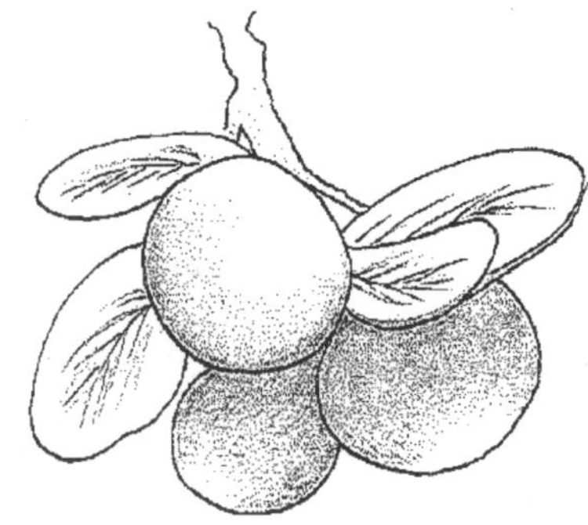
Kei apple

Thorny small evergreen tree with yellow-orange, apple-like fruit that is tart (acidic) to eat (you have to add sugar). Has male and female plants. Frost and drought tolerant. Contains high vitamin C (more than oranges) and a good range of amino acids and proteins (high in pectin too). Leaves used as fodder, fruit also eaten by animals. Used to make jams and jellies. Although from subtropical Africa it survives well in drier areas.

We should also mention Natal plum (*Carissa macrocarpa*), which is another thorny shrub from South Africa, as an alternative to Kei apple. The small crimson fruit can be eaten whole, and is very high in vitamin C, and has fair amounts of iron, potassium and copper.

This salt-tolerant, evergreen plant is well adapted to dry coastal areas and is both drought and cold tolerant, but does prefer warmer, moist climates for its best production.

Marula (*Sclerocarya birrea*)

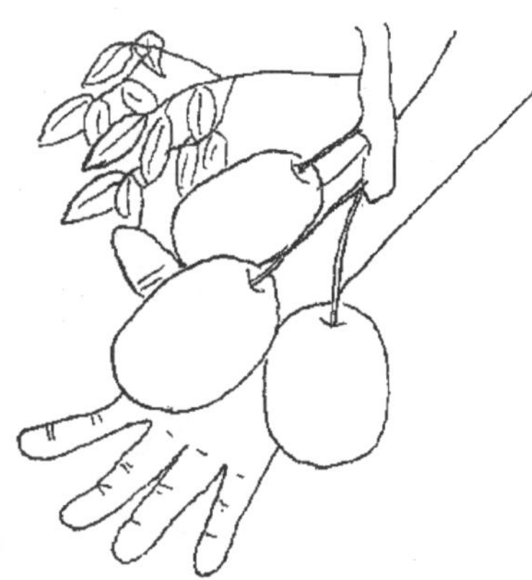
Marula

A large-spreading, deciduous tree from Africa, this is a multipurpose plant. The fruit, seeds, leaves and flowers are all edible and the bark is used traditionally for medicinal purposes.

The fruit contains four times the vitamin C of oranges. However, it is the kernels inside the central seed that are most valued by humans, although the fruit can be made into jam and fermented into a liqueur. Animals have been known to become drunk when they eat the fallen, fermented fruit. The kernels are high in protein, magnesium, potassium, phosphorus and fat. Oil from the kernel has antioxidant properties and is used for cooking and as a moisturizer for the skin.

Olive (*Olea europaea*)

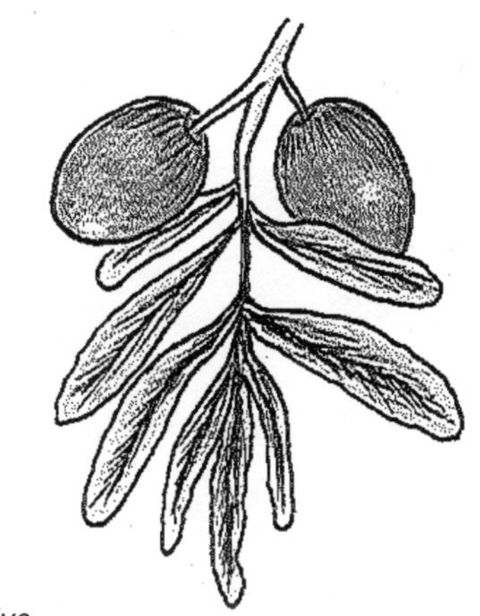
Olive

Nutritionally, olives are not that great. While they do have reasonable levels of iron and fiber they are often also high in sodium (salt) and lacking many vitamins and minerals. Despite this, olives have been cultivated for over 5,000 years as we just love to cook and drape food with their oil. Olive oil is cholesterol-free and mainly contains both mono- and polyunsaturated fatty acids — so it is generally healthy to consume.

The fruit need to be pickled to enable them to be eaten. They survive in dry climates because they have extensive root systems, but need to receive over 20 in annual rainfall to fruit well.

Pistachio (*Pistacia vera*)

Pistachio

Pistachios are deciduous trees growing to 15 ft. The plant is not self-fertile — there are male and female plants. The seed (nut) is rich in edible oil and it has high levels of fiber, protein, vitamin B6, thiamine, phosphorus, copper and manganese. Pistachio nuts are widely used in confectionery, ice cream and cakes in addition to their main use as a snack nut.

Pomegranate (*Punica granatum*)

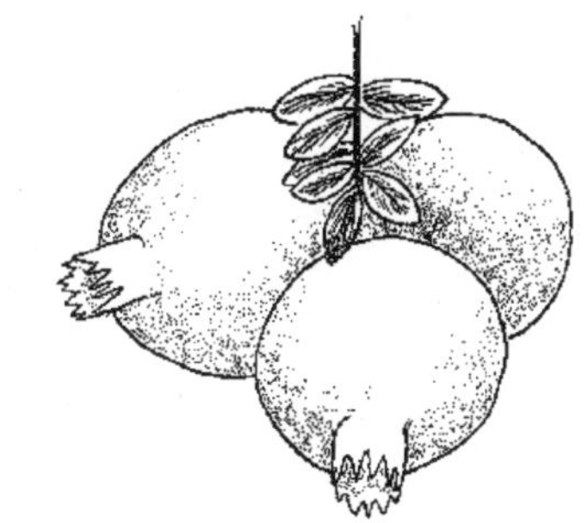
Pomegranate

Originally from the Middle East, the pomegranate is a small deciduous tree, the fruit of which is high in fiber, folate, vitamins C and K, and minerals potassium, copper and manganese.

The fruits split open when ripe and the pulpy seeds are often sucked and then spat out, which is a shame as the seeds are edible and contain lots of nutrients. Pomegranates are more often consumed as a juice, and they are used in soups and beverages such as grenadine.

Cultivation notes

Many other fruiting plants can be grown in dry regions. Grapes, mulberry, Natal plum, peach and apricot can survive provided additional watering is undertaken.

Fruit or nut tree	Cultivation
Almond	While you can buy self-fertile varieties, most almonds need a pollinizer tree nearby to enhance pollination.
Apple cactus	Don't over-water (if at all, depending on your climate).
Dragon fruit	Provide a frame or trellis on which they can climb and spread out. They are originally from the subtropics, but seem to do well in drier climates.
Figs	Keep restricted to a manageable size, otherwise they just keep growing. Prune after leaf fall.
Jujube	Buy grafted varieties as plants grown from seed or cuttings never seem to produce large fruit and they may take many years before they produce.
Kei apple	Need male and female plants so get at least three plants to increase your chances of having both sexes. It is extremely thorny and makes a great living fence or barrier. Handle with care.
Marula	Can get large, so plant where you have room. There are separate male and female plants, so if you want fruit buy a few plants and hope for the best.
Olive	Choose the type you want — either for table (eating) or for oil. Some varieties have both functions. Some you pick when green, others when black, and others either color.
Pistachio	Obtain at least one male plant for every ten female plants.
Pomegranate	Can be kept in a large pot. It can become scrappy, so when it is dormant prune dead wood out and internal twigs and branches to open up the plant in the center.

Table 6.1. Cultivation notes for dryland fruits and nuts.

» DID YOU KNOW?

Many flowers are dioecious — individual flowers are either male or female, but only one sex is to be found on any one plant, so both male and female plants must be grown if seed is required. These plants benefit when two different varieties are planted close to each other. Plants in this category include carob, pistachio and kiwifruit. Monecious plants are those that have male and female flowers or reproductive parts on the same flower or plant. Most plants fall into this category.

Cold climate fruit and nut trees

Cold climate areas have annual rainfalls from 24 to 40 inches or so, and include temperate and Mediterranean climates.

While our top ten ranking of fruits is based upon their content of various nutrients such as vitamins and minerals, different texts will list varying levels of these. So ultimately there will be conflicts and differences of opinion about what should be on the list.

Availability, ease of growing and harvest, storage, other uses and their survival in a range of different soils and climates also influenced this list.

Apricot (*Prunus armenica*)

Apricots are a good source of vitamins A and C, fiber and potassium. They are very low in saturated fat, cholesterol, starch and sodium.

Apricots make a good dried fruit, and if you are prepared to crack open the "stone" you can press oil from the seeds (or kernel) or eat them as a substitute for almonds. A medium-sized deciduous tree (all stone fruit are).

Apricots will grow in dry climate areas provided they experience enough chill hours.

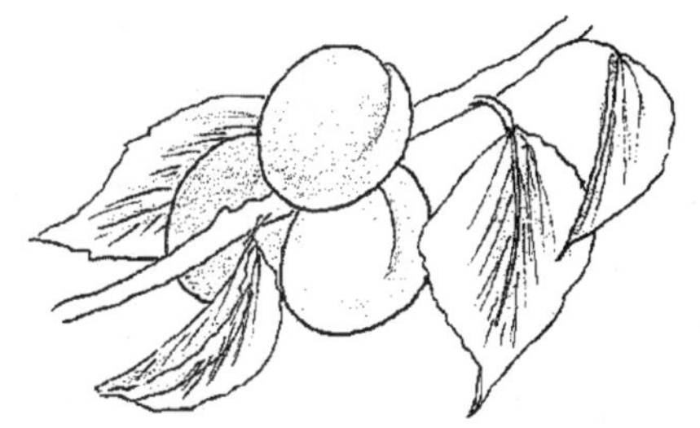

Apricot

Grapefruit — pink or red (*Citrus paradisi*)

Very good levels of vitamins A and C, and potassium and fiber, and smaller amounts of most other vitamins and minerals.

The red varieties are sweeter than the normal variety, so most people should be able to eat these raw. An evergreen small tree or large shrub, grapefruit have "wings" at the base of the leaves, and this is how you can distinguish them from other citrus.

If you don't think you would eat a grapefruit then plant a lemon tree. Lemons are more cold tolerant than grapefruit.

Grapefruit

Hazelnut (*Corylus avellana*)

The hazelnut is a deciduous tree that grows to 20 ft × 10 ft. It is easy to grow, produces suckers and yields in 3–4 years. They need to be protected from strong winds, certainly in their early life.

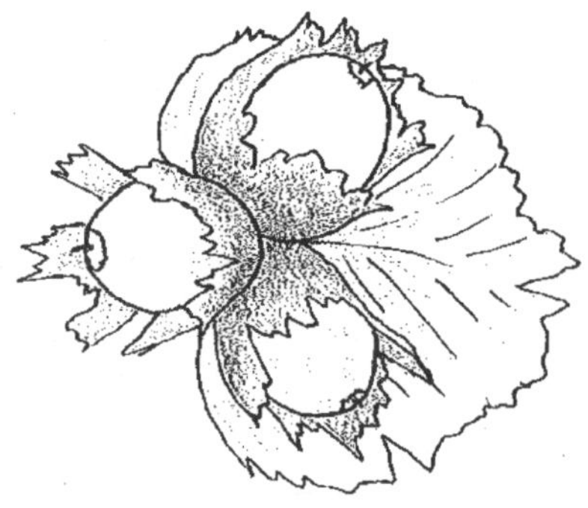

Hazelnuts

Hazelnuts are wind pollinated, and require another variety nearby for pollination. A mature tree (10 years) may yield between 15 and 20 lb of nuts. Hazelnut trees have additional value as hosts for truffles, which can be lucrative.

Hazelnuts are very high in fiber, protein, monounsaturated fat, vitamin E, thiamine, magnesium, manganese and copper.

» DID YOU KNOW?

Many fruit trees can be espaliered or kept in large tubs, if space is limited. You can train and trellis apple, persimmon, apricot and pomegranate. Passionfruit and kiwifruit are vines anyway. Most types of citrus, guava, apples and hazelnut can be placed in containers, and moved around as required.

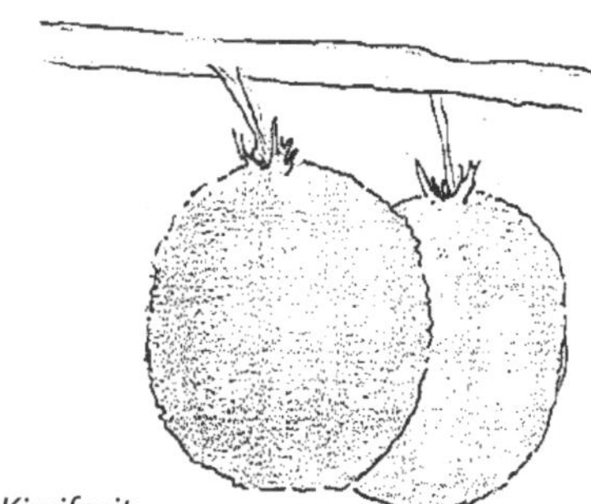

Kiwifruit

Kiwifruit (*Actinidia deliciosa*)

Kiwifruit (also known as Chinese gooseberry, where they originated) are high in vitamin C, fiber, potassium and magnesium.

You need male and female vines; they tolerate cold and you grow them on a well-supported trellis as they can be prolific producers. They like sun but will not survive long summer weather. Deciduous vines.

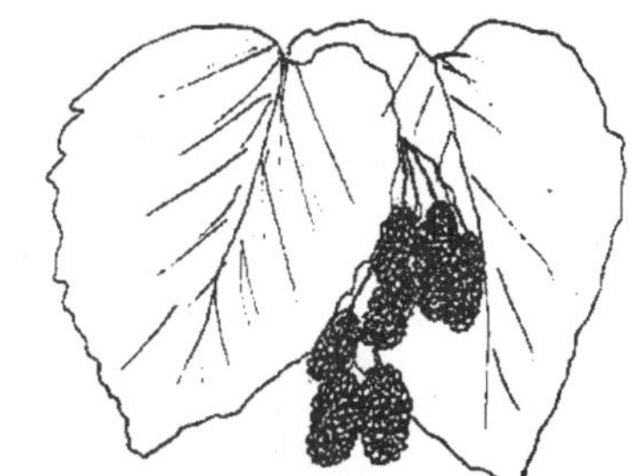

Mulberry

Mulberry (*Morus nigra*)

There are many cultivars and species of *Morus,* which can hybridize. The black mulberry is the most common, but red and white mulberries are becoming popular too.

Mulberries are high in vitamins C and K, iron, magnesium and potassium. They are large deciduous trees with extensive root systems.

Mulberries can grow in drier climate areas provided their roots can tap into underground water supplies.

Orange

Orange (*Citrus sinensis*)

These are sweet oranges and cultivars include seedless varieties. An evergreen small tree, oranges are high in vitamin C and have fair amounts of vitamin K, riboflavin, fiber, iron and potassium.

The bitter orange (*Citrus aurantium*), also known as Seville orange, is commonly used to make marmalade.

Passionfruit (*Passiflora edulis*)

An underrated food plant, although it can be difficult to grow and keep alive. Some vines are aggressive and grow forever, while others just die for no apparent reason.

Vines grown from seed are usually short lived, so known passionfruit varieties are often grafted onto vigorous, long-lived rootstocks. Unfortunately, these rootstocks tend to sucker and spread throughout the garden.

Passionfruit are high in fiber and protein, vitamins A, C, riboflavin and thiamine, and minerals iron, potassium and magnesium. This evergreen vine can spread over a trellis, wire fence or chicken pen enclosure.

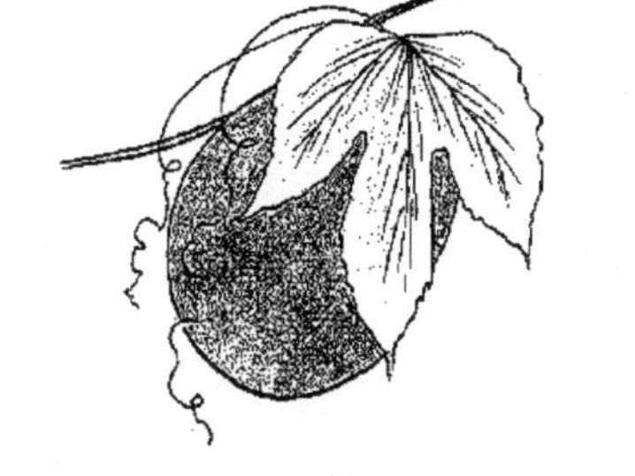

Passionfruit

Pecan (*Carya illinoinensis*)

Exceptionally high in manganese and very high in fiber, protein, thiamine, vitamin B6, magnesium, phosphorus, zinc and copper, pecan nuts have no cholesterol even though they do have your daily allowance of fats in a cupful (the fats are mainly mono- and polyunsaturated).

Pecan trees can get large, so you need a big backyard or a paddock to allow them to spread. Like almonds, pecans are not true nuts but are a drupe — a single "stone" surrounded by flesh. In this case, it is a seed inside a thin shell, inside a husk. The green husk matures and dries to brown and splits open.

Pecan

These long-lived, deciduous trees usually need another nearby for pollination. I should mention that walnuts have similar nutrient levels (even higher in some nutrients) but they require specific growing conditions and a narrow climate. Walnuts are big trees too but they may be more suitable for your area.

Persimmon (*Diospyros kaki*)

The persimmon is a deciduous small tree, and different varieties produce fruit that can be astringent or non-astringent.

Persimmon

This means that they either taste sandpapery due to excessive tannins in their fruit or they are firm and sweet.

The most popular persimmons are the non-astringent type. However, when astringent varieties fully ripen they are very sweet as well.

Persimmons have good quantities of fiber, vitamins A and C, and minerals manganese and potassium. They are usually picked when they are an orange-red color, still firm and allowed to ripen off the tree. It is a misconception that they should be left to really soften and be almost rotten.

Strawberries (*Fragaria ananassa*)

Strawberries

It is hard to imagine that strawberries have more vitamin C than oranges, but this is the case. Besides vitamin C, strawberries have fair amounts of fiber, folate, manganese and potassium.

They also have very little fat and cholesterol, and surprisingly, relatively low sugar.

Strawberries are easy to grow and propagate for the next season. Strawberry plants are essentially low-lying, evergreen herbs.

Blueberries, lemon, feijoa, loquats, walnuts and grapes all should get a worthy mention in the next group of nutritious fruits.

Very common fruits such as apples, pears, peaches and plums, and less common fruits such as quince and limes, are nutritious but a little further down the list.

» DID YOU KNOW?

Strawberries are not true fruit either. A fruit is defined as "a ripe ovary and its seeds", and this includes oranges, passionfruit, apricots and persimmon.

Strawberries are formed from a swollen receptacle, which is the section below a normal ovary. The "seeds" seen on the outside of a strawberry are the ovaries and the seeds are inside these.

Apples and pears are also false fruits. The core of both the apple and pear is the ovary (seeds inside). We eat the swollen, juicy receptacle, and throw away (usually) the ovary. For normal fruit we eat the female reproductive organ of the plant: what a thought!

Cultivation notes

Fruit or nut tree	Cultivation
Apricot	Can get fruit fly attack in some areas.
Grapefruit	While not that popular, they are tough and hardy. The fruit can interfere with some types of medications, especially those taken for any cardiovascular complaints (cholesterol, heart, blood pressure etc.).
Hazelnut	A good understory crop for oaks and pecan nut trees.
Kiwifruit	Make sure you obtain male and female plants. They don't tolerate full sun and drying wind early in their life. Protect them from frost.
Mulberry	Maintain their height as they want to grow large. Some red and white varieties can be smaller.
Orange	Orange trees love full sun, so plant a citrus grove in a warm, open position. You might like to try a red (blood) orange or a Seville for variety.
Passionfruit	Heavy feeders, so add manure and compost every now and again.
Pecan	These want to grow up to be giants, so make sure you have the space. Many pecan varieties named after Native American tribes. Seem to do better with neighbors nearby for pollination.
Persimmon	Can be espaliered along a fence. If you leave the fruit to mature as long as you can (without it going too soft and rotten) the tannins slowly diminish and the sugar levels rise. The fruit is better to eat at this stage.
Strawberry	Strawberries generally have male and female flowers, but are normally propagated by their runners. Keep the developing fruit off the ground and away from snails and slugs.

Table 6.2. Cultivation notes for cold climate fruit and nut trees.

Warm humid climate fruit and nut trees

Humid climates are the tropical and subtropical areas of the world, with annual rainfall of 30 in and more, but most often well over 40 in.

While most westerners would call these exotic fruit, indigenous peoples living in the warmer and humid areas of the world just know them as staples.

Avocado (*Persea americana*)

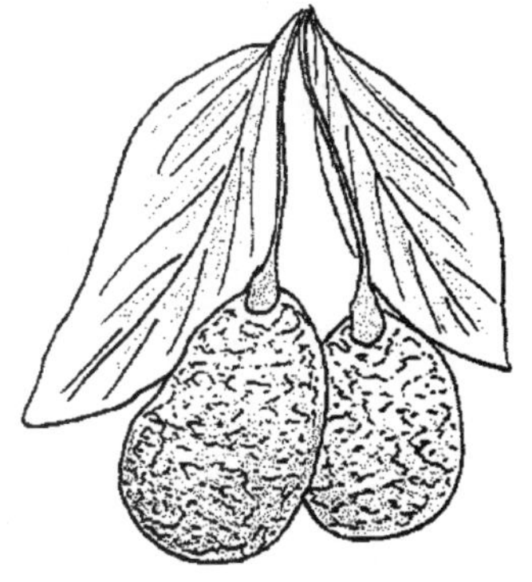

Avocado

Avocado are tall evergreen trees that produce a berry fruit with a single large seed. The fruit is a good source of vitamins C and K, folate, potassium and fiber. Well known as a "fatty" fruit, the fats are mainly mono- and polyunsaturated fats and so can be added to diets where meat or dairy products are not available. Fruits are picked green and allowed to ripen off the tree.

Avocado trees do not tolerate frost well, although some varieties are cold-adapted. Plant them in warm, sheltered areas away from strong winds.

There are two types, A and B, each producing male or female flowers at different times of the day. Obtain both types if you can as most avocado trees are not self-fertile and rely on another tree nearby for pollination. Certainly production will be greater with both varieties, but in a domestic situation a semi self-fertile variety will produce enough fruit for the family.

Banana (*Musa* spp.)

Banana

Bananas are another fruit that is seldom allowed to mature on the plant but are picked green and ripened off the plant. Bananas can be more than 10 feet tall, but they are not trees, they are very large herbs. Their stems are layers of leaves tightly wrapped around each other.

The fruit is high in vitamins B6 and C, potassium, manganese and fiber, but also contains a high sugar content. Sweet types are just "bananas," while starchy types used in cooking are plantains. There are many cultivars and debates about scientific naming.

Some varieties can be grown in temperate climates but unless they are kept in a warm, protected area they will not produce much fruit. Bananas need consistent soil moisture on very fertile soils to produce good crops.

Brazil nut (*Bertholletia excelsa*)

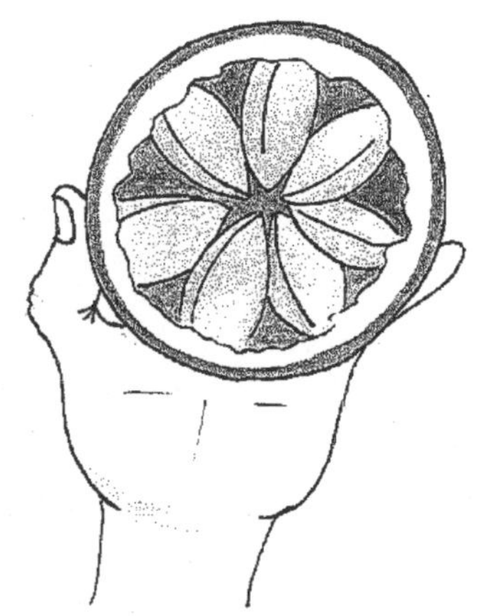

Brazil nut

While Brazil nuts are high in saturated fat, they do contain exceptional levels of selenium, magnesium, copper, phosphorus and manganese, with very good amounts of fiber, vitamin E, thiamine and zinc. Like many foods, small amounts in your diet should be beneficial.

Brazil nut trees are very tall, semi-deciduous (may lose leaves in dry season) and produce very large fruit containing many seeds — the nuts. There is no opportunity to plant these trees outside of South America as they rely on a rainforest orchid and particular orchid bees for pollination: a classic example of symbiosis.

Cashew (*Anacardium occidentale*)

Cashews are nutritious nuts that are good sources of omega-3 and omega-6 fatty acids, protein, vitamin K, vitamin B6, iron, magnesium, phosphorus, copper and manganese.

Cashew nut trees are evergreen and medium sized, and often spread as much as they are tall. The cashew nut is the seed of a large fruit, which is also edible and can be made into drinks. The cashew apple, as the fruit is known, is another false fruit, because it arises from a swollen receptacle at the base of the ovary. The ovary and cashew nut are found external to this, like a prune hanging on the end of the fruit.

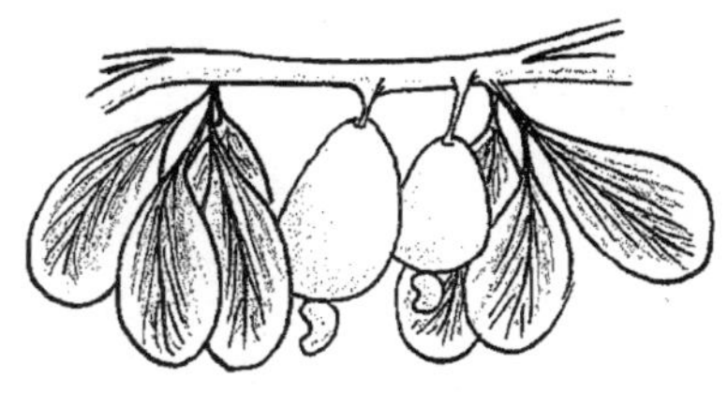

Cashew

The cashew seeds are found inside a tough shell full of caustic oil, so care is required during harvest. The nuts are normally roasted or deep fried to remove any traces of the chemicals, which will cause discomfort if the nuts are eaten raw.

Guava (*Psidium guajava*)

Guava has the highest concentration of vitamin C amongst the common foods, although the Australian bush food Kakadu plum has more (it has 3,000 mg/100 g fruit, which is 12 times more than guava). Besides vitamin C, guava has fair amounts of fiber, vitamin A, folate, potassium, copper and manganese.

Guava

An evergreen shrub, the common guava is also known as the apple guava. The strawberry guava (*Psidium littorale*) has fewer nutrients but still enough vitamin C to meet the daily requirement after one helping of fruit. The common guava can survive in deserts too, provided they get occasional water.

Mamey sapote (*Pouteria sapota*)

"Sapote" is an Aztec term meaning soft, sweet fruit, so it is given to three common fruits that are all unrelated. The little-known mamey sapote from Central America is completely different from the black sapote and the white sapote.

The mamey sapote is a huge evergreen tree that produces large oval-shaped, pink fruit with a brown skin. The fruit is an excellent

Mamey sapote

source of vitamins B6 and C, and has fair levels of vitamin E, manganese and potassium. The fruit can be eaten raw or made into drinks and jam.

The black sapote, *Diospyros digna,* is related to persimmon, and while nutritious with fair amounts of vitamins A and C, fiber and potassium, these are typically lower than that found in mamey sapotes. The white sapote, *Casimiroa edulis,* is more related to citrus, and is a cousin of the custard apple. It appears to be less nutritious still, although its seeds are being investigated for possible medicinal purposes.

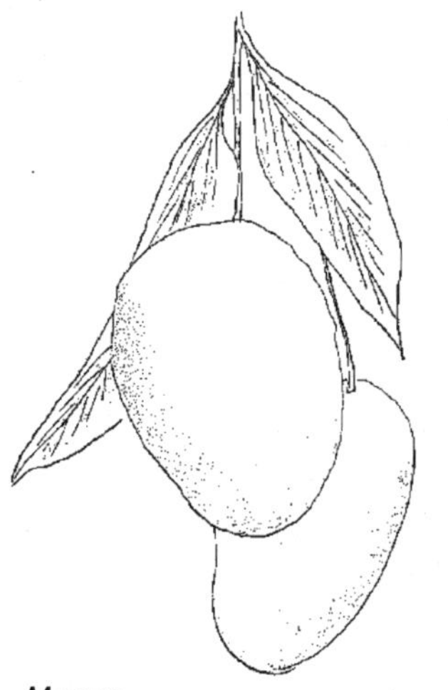

Mango

Mango (*Mangifera indica*)

Long-lived, evergreen trees that can grow to 100 ft, although commercially they are kept to a picking height less than 12 ft. Another tree that can grow in temperate climates and these varieties are normally grafted onto cold-hardy rootstock.

Some people refer to mangoes as the "king of fruits" because they are so delicious. The fruit is high in fiber, vitamins A, B6 and C, and contains fair amounts of folate, potassium and copper.

Papaya (*Carica papaya*)

Papaya (more commonly known as pawpaw in Australia, Africa and the UK) is very high in vitamin C and has good amounts of vitamin A, fiber, folate and potassium. There are male and female flowers on different plants so you need a few plants to ensure fruit will develop on at least one of them.

Some self-fertile (hermaphrodite) and cold-adapted varieties are common commercial plants and are now starting to emerge in nurseries. Papaya like warm sheltered spots, so if you can provide the right microclimate you will have success in cooler regions. Seeds germinate relatively easily and you need to plant papaya in well-draining soil as prolonged waterlogging will kill the tree.

Different varieties will produce yellow, orange or red flesh. The seeds are also edible and can be ground to make a substitute for black pepper.

Papaya

Pineapple (*Ananas comosus*)

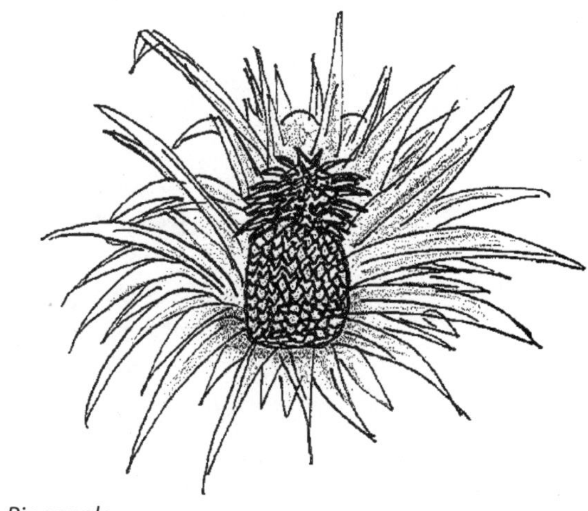
Pineapple

High in vitamins B6 and C, thiamine and manganese, pineapples are easy to grow from a tip cutting. Pineapple is virtually fat-free, cholesterol-free, salt-free and starch-free. The fruit is consumed fresh, cooked or juiced, and it contains enzymes that break down protein, so it tenderises meat and is a useful component of a marinade.

The plant is typically only about 3 ft high and can be kept in a large tub, even indoors. Unfortunately it takes about two years to get mature fruit so you have to be patient.

Soursop (*Annona muricata*)

Soursop

Soursop, also known as graviola, is a member of the custard apple family. It is a small (12 ft) evergreen tree that produces fruit rich in vitamin C, thiamine, riboflavin, niacin, iron, magnesium and potassium.

Like most tropical plants it cannot tolerate frost and cold winds. Besides eating the fruit fresh, it can be made into juice, fruit leather and used in cooked meals.

As you would expect, two closely-related species — the cherimoya (*Annona cherimola*) and custard apple (*Annona reticulata*) — have nutritious fruits with good levels of vitamins C and B6, potassium and riboflavin.

There are lots of other tropical and subtropical nutritious fruits and the next wave would include longan, mangosteen, coconut, jackfruit, acai palm and star fruit.

» DID YOU KNOW?

Lemon juice prevents oxidation, so it can be poured over apple and avocado slices to prevent them from turning brown.

Cultivation notes

Fruit or nut tree	Cultivation
Avocado	Fruits only ripen after harvest, so they can be left on the tree for some time, and picked as you require them. Each flower opens twice over two days, firstly as a female and then as a male on the second day. Avocados need free-draining soils and a good shovelful of compost every month when they are producing fruit.
Banana	Once the banana bunch has been picked, that "mother" stem should be cut down. Each stem only ever produces a single bunch. New suckers emerge to make future fruit. You might choose to keep the large, strongest suckers (maybe two) and remove the rest. The plants are heavy feeders so fertilize every few months.
Brazil nuts	These trees do not produce nuts in cultivation outside the rainforests of South America, so your only source is from bought ones. One of the few trees that are harvested in their natural habitat.
Cashew	Easy to germinate fresh seed, and grow and maintain plants. The biggest problem is removing the nuts from their shells without coming into contact with the oil, which can burn the skin and severely irritate eyes.
Guava	Reasonably drought tolerant for a tropical plant, and survives in poor soil. Small tree or large shrub, fruit vary as with the cultivar. Whole fruit can be eaten, although most people just eat the flesh and discard the skin. Can be prone to fruit fly, so hang baits to minimize damage.
Mamey sapote	Majestic trees that need space to grow. Like most rainforest trees they require well-drained, rich soils, free from frost and damaging winds. You can prune to a required height to enable easier picking and general maintenance.
Mango	Some varieties produce seeds that are polyembryonic — several seedlings grow out of one seed, which are usually true to type (no need to have a grafted variety to produce good fruit). Choose the largest, healthiest seedling and carefully remove the others. Self-pollinated, so fruit will form in a single tree.
Papaya	Usually grows as a single-stemmed tree, with no branches and just a crown of leaves at the top. The stem is soft and can rot, causing the plant to topple over. Seed is easy to germinate in moist, rich (compost) soil. As long as you keep seedlings protected from too much sun and wind, and sheltered from the cold, you will have great success.
Pineapple	Cut the crown off a pineapple. Remove most of the outer leaves and leave the (cutting) stalk for a day or two to "heal" the wound. Place cutting in a flowerpot full of good propagating mix, and leave in a warm position (hothouse is ideal). Don't over-water (the roots rot easily), and when you see new growth, it's time to plant in the garden. The plants are prickly so keep them about 5 to 7 ft apart.
Soursop	Can be grown from seed and fruit is often true to type. These trees are hard to keep alive as they shut down in cold weather and hate getting "wet feet." They must be planted in free-draining, sandy soils and in warm, sheltered positions. If they get too hot or too cold they drop their leaves, and you might not be able to save them.

Table 6.3. Cultivation notes for warm humid climate fruit and nut trees.

Some other considerations

Some types of fruit trees may not produce fruit unless they are subject to cold temperatures over a period of time. Stone fruit, such as cherries, plums and peaches, as well as pistachios, walnuts, apples and pears, for example, require a certain number of "chill" hours. This refers to the number of hours below 45°F that the trees are subject to, as they need a certain amount of "cold" to trigger the development of leaf and flower buds (and therefore fruit). If the chill hour requirements are not met, the plant will develop leaves sporadically over the tree, flowering will be disrupted and little or no fruit will be produced. For this reason, selecting varieties of fruit trees that match the chilling hours for the area is essential for successful fruit production.

If you live in a warmer area then look for varieties with a low chilling requirement. These fruit trees produce fruit "early" in the season. Early, low chill varieties are a useful strategy to miss fruit fly and other pests that attack summer crops. For example, low chill peaches and nectarines may produce fruit in May and June. The fruit fly season starts in July and goes through to September. This means that you will be able to have a good feed of organically grown fruit well before the fruit fly is about.

» DID YOU KNOW?

Fresh fruit is always the best to eat. Heating, canning and drying any fruit depletes vitamins, enzymes and a host of other organic substances. Heat and other processes denature (rearrange and break down) molecules so they become ineffective. Canned fruit often has its fiber-rich peel removed and some fruits are packed in syrup, which contains high levels of sugar. So up go the calories.

Size does matter! Many fruit trees can be found and bought in one of three forms: dwarf, semi-dwarf or standard. Standards are the old traditional fruit trees. These need to be regularly pruned or they become very large trees.

Dwarf fruit trees are small trees for small spaces, and they are becoming more popular for the home gardener because they produce roughly the same amount of fruit for the space they take up as do standard trees, and at a younger age.

Harvesting and pruning are both easier because of their size. You can reach all parts of the tree from the ground without using a ladder.

Dwarf varieties are often better to espalier than standard varieties. However, some dwarfs are not as sturdy in high winds and need support when they are heavy with fruit, and many also don't live as long as standards.

» HANDY HINT

Citrus leaves. You can tell the species by the shape of the leaves.

L to R: Grapefruit has small wings at the base of the leaf. Mandarin has long, narrow leaves. Orange and lemon have squatter leaves, and a little rounder.

Crush and smell the leaves and you should be able to distinguish between a lemon and an orange.

HEAVENLY HERBS 7

HERBS ARE SMALL PLANTS that have useful functions or useful products such as edible leaves, seeds or flowers, scented leaves for pest control, high levels of nutritious vitamins and minerals, aromas and flavor to spice meals, essential oils, medicinal uses and a myriad of other uses for our health and wellbeing.

It is easy to see why permaculture designers incorporate a range of herbs into their designs, acknowledging the important role herbs play in cultivated ecologies.

However, it is extremely difficult to pick a "top ten" of all the herbs we use and value, so I have opted for the top ones in different categories of herbs: edible, pest-repellent, medicinal and herbal teas.

Herbs used in cooking

Usually it is the leaves that are used in cooking and most herbs used in this way, such as rosemary and fennel, add a savory flavor to food.

Some herbs, such as tarragon and basil, do add a mild sweet flavor, but the majority of herbs are either sour or bitter. It is the oils and other aromatic chemicals in the leaves of herbs that provide the seasoning and flavors to food.

When other parts of plants are used, such as the seeds, bark or root, to provide flavoring or coloring to food, we call them spices. Spices include turmeric, cinnamon, ginger, cloves, galangal and nutmeg.

Basil (*Ocimum basilicum*)

Sweet basil is the common garden variety that is traditionally used in many tomato and meat dishes, and in making pesto. Basil leaves do have a fair amount of vitamin K but not much else. The herb is used fresh in cooking as drying destroys much of the flavor.

Sweet basil

Many other varieties are used in cooking dishes and these include Thai, holy, Greek and lemon basil.

Most basil types are annuals (although in warmer climates basil may survive through winter) but you can buy a perennial basil, which can be pruned and the leaves used as required.

While the leaves of perennial basil are not used for salads, they can be added to cooked vegetable and meat dishes.

Bay leaf

Bay leaf or bay laurel (*Laurus nobilis*)

Bay leaves are used to flavor soups, meat dishes and sauces, and are a key ingredient in bouquet garni and chai tea.

While fresh leaves can be used, bay leaves develop their full flavor a week or two after picking or drying. Bay leaves are very tough even after cooking and are normally removed from the dish before eating. They are not particularly nutritious.

» DID YOU KNOW?

Many herbs can be dried by hanging bunches in a dark, cool, well-ventilated place for a few weeks. While it is usually best to use fresh herbs, dried herbs such as bay leaf, lavender, thyme, mint and sage, still maintain some of their smell and flavor for more than a year.

There are many other trees similar to bay laurel, where the leaves are used in much the same way. These include the Californian bay tree (*Umbellularia californica*) and Indian bay leaf (*Cinnamomum tamala*).

Chives

Chives (*Allium schoenoprasum*)

Unlike most of the onion family, chives are a small, clumping, perennial plant and while it does have a little bulb, it is the chopped leaves that serve as a garnish and are added to soups and egg, vegetable and meat dishes.

Chives were originally found across the northern hemisphere, but are now cultivated throughout the world. In colder climates chives die back to the bulb, but sprout again when warmer weather appears.

The leaves contain considerable amounts of vitamins K, C and A, and the minerals manganese and magnesium.

Garlic chives (*Allium tuberosum*), a similar species that has strappy leaves, are used in Asian cuisine.

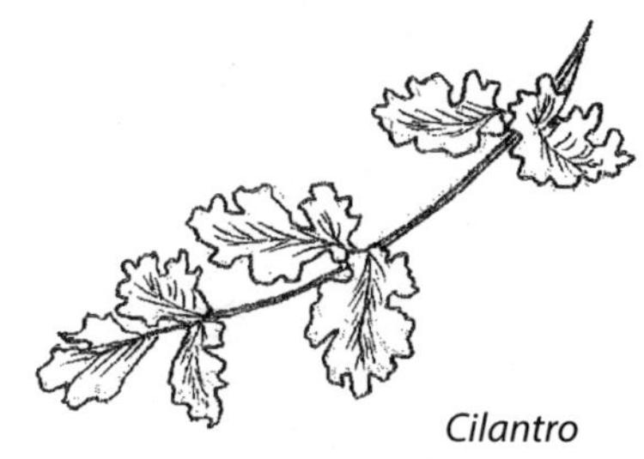
Cilantro

Cilantro (*Coriandrum sativum*)

This is an annual herb growing to about 20 in tall. Fresh leaves are added to chutneys, salads and curries, and often as a garnish on cooked food. The crushed root is used in Thai cuisine. The leaves contain high amounts of vitamins K, C, A and E (in this order) and reasonable amounts of the minerals iron, manganese and potassium.

The seeds (dry fruits) can be used whole or ground to a powder, and this spice adds an aromatic, citrus flavor to many dishes. Cilantro seed is also called coriander.

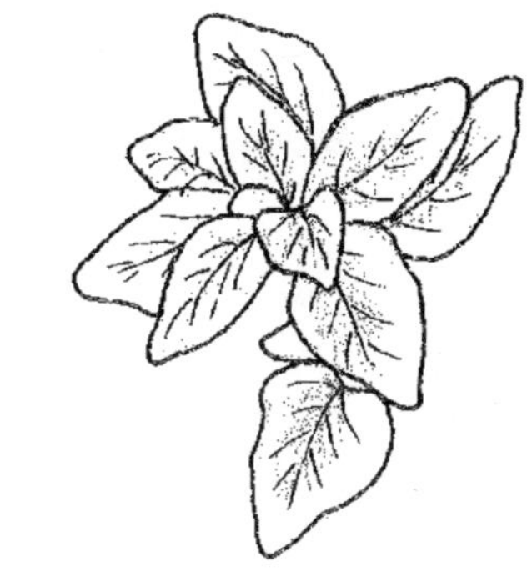
Oregano

Oregano (*Origanum vulgare*)

Common oregano is a small, perennial herb to 20 in, although in colder climates it can die off in winter. Like bay leaves, the flavor of the dried leaves is strong. However, both fresh and dried leaves are added to pizzas, tomato and meat dishes, salads and roast vegetables.

High in fiber, calcium, iron and magnesium, oregano leaves also contain many organic compounds, which vary in concentration depending on the variety or cultivar grown.

A very closely-related species, marjoram (*Origanum majorana*), is used like (and confused with) oregano, but it has a slightly sweeter, gentle flavor. While marjoram is used in French cooking, Italians prefer oregano.

Curly-leaf parsley

Parsley (*Petroselinum crispum*)

Parsley is high in vitamins A, C and K and is a common garnish for salads, meat and potato dishes, and soups. The curly-leaf variety (var. *crispum*) is more decorative than the flat-leaf variety (Italian, var. *neapolitanum*), but both are easy to grow from seed. In cooler climates parsley may be biennial but in warmer climates it tends to be an annual.

Flat-leaf parsley or Italian parsley is larger than the common curly-leaf variety.

Rosemary

Rosemary (*Rosmarinus officinalis*)

Rosemary is a drought-tolerant, tough, woody perennial to 3–6 ft high that has a long tradition in cooking. Often a sprig is used on top or inside a roast of poultry, red meat or fish, while a few finely-chopped leaves add a tasty garnish to bread, pizza or other pastries.

A sprig or two placed in olive oil for a few weeks causes the essential oils in the rosemary leaves to permeate the oil. The leaves themselves are not nutritious but they do add a lovely favor to dishes.

Summer savory

Summer savory (*Satureja hortensis*)

Summer savory is a small annual plant, growing up to 20 in tall, that is easily grown in a container, and the dried or fresh leaves flavor vinegars, soups and meat dishes. While the leaves do contain fair amounts of manganese, calcium and iron, summer savory is much better known for the peppery tang it adds to foods.

» DID YOU KNOW?

There are many species that belong to the "savory" family. For example, you can also use winter savory (*Satureja montana*), which is a perennial plant growing up to 12 in tall, and has a similar spicy flavor (although a little more bitter) to summer savory.

Tarragon

Tarragon (*Artemisia dracunculus*)

The most commonly used culinary variety is French tarragon. This is an aromatic, perennial herb and is used to flavor sauces, meat dishes and the occasional drink. Nutritionally very poor, tarragon oil contains several organic compounds that give it a particular anise taste.

Thyme (*Thymus* spp.)

The tiny leaves of many thyme species are added as seasoning to omelettes or scrambled eggs and a variety of meats and sauces. Like rosemary, finely chopped or dried thyme leaves can be sprinkled onto bread dough to make a savory loaf. For a mild citrus flavor use lemon thyme (*Thymus citriodorus*). The leaves do not contain many nutrients, but do contain an aromatic compound called thymol that is a good antiseptic and gives thyme its strong flavor.

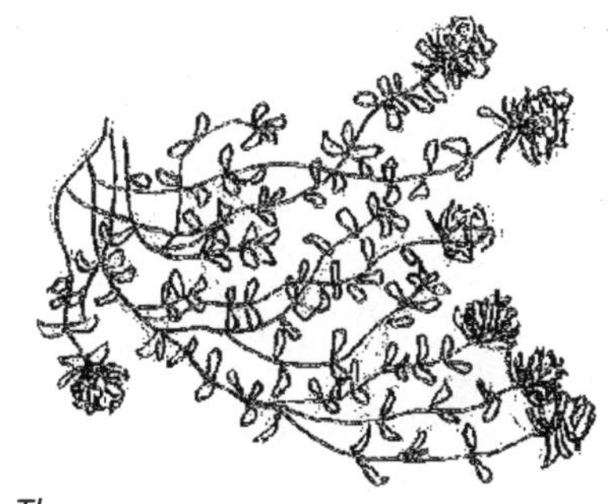

Thyme

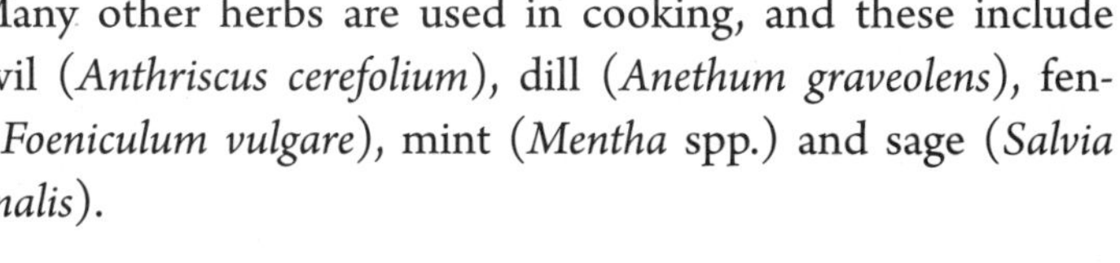

Many other herbs are used in cooking, and these include chervil (*Anthriscus cerefolium*), dill (*Anethum graveolens*), fennel (*Foeniculum vulgare*), mint (*Mentha* spp.) and sage (*Salvia officinalis*).

Herbal teas

Herbal teas are soothing, usually free from caffeine and very satisfying, fresh from your garden. Usually, three to six leaves (about one sprig or two teaspoons fresh, one teaspoon dried) are infused in boiling water for a few minutes — leave longer in the cup for a stronger flavor. You can sweeten with honey or add a slice of lemon or ginger. If you were keeping the tea for a drink later (placing it in the refrigerator), then strain to remove the leaves.

Some herbs can be used together, such as green (Chinese) tea and bergamot, lemon verbena and cinnamon, and bergamot and lemon verbena.

Bergamot

Bergamot (*Monarda didyma*)

This perennial herb is also known as crimson beebalm, and it grows to 3 ft high. It is a hardy plant with orange-scented foliage that may help mask susceptible plants from pests. The flowers attract bees, birds and butterflies. Aromatic teas are made from fresh or dried leaves and flower heads.

Dandelion

Dandelion (*Taraxacum officinale*)

The dried and roasted root is ground to make a substitute for coffee. The edible leaves are high in vitamins K, A and C, and the minerals iron and manganese, and can be added to salads (although used in moderation). Older leaves are usually cooked as a vegetable.

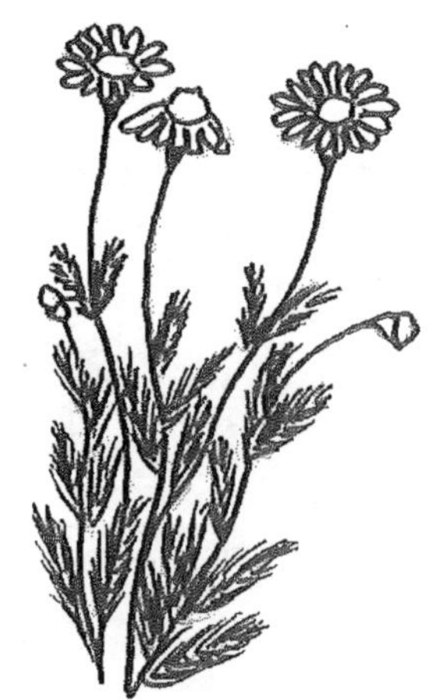

English chamomile

German chamomile

Ginger

Other herbs can be mistaken for dandelion. Dandelion has a single flower stem while similar-looking herbs, such as catsears and hawksbeard, have multiple flowering stems.

» DID YOU KNOW?

In botany (and cooking), a "sprig" is a small piece of stem or branch with leaves attached. It is usually about 2–3 in in length but there is no exact defined length. You might cut a sprig of rosemary to add to a roast meat dish or a sprig of lavender placed in a clothes drawer to repel silverfish.

English or Roman chamomile (*Chamaemelum nobile*)

Use a teaspoon of dried flowers in a cup of boiling water. English chamomile is not as sweet as German chamomile. It is useful as a perennial herb lawn because it is a creeping ground-cover plant to 12 in high, and its leaves are finely divided and aromatic.

German chamomile (*Matricaria recutita*) [syn. *M. chamomilla, Chamomilla recutita*]

German chamomile is a low annual shrub with daisy-like flowers similar to English chamomile. Use fresh or dried flower heads to make tea. It is reputed to be good for calming nerves.

Ginger (*Zingiber officinale*)

Ginger requires specific growing conditions as it is a tropical plant that requires rich soil, a warm, sheltered area and lots of water.

Grate or slice about half a teaspoon of ginger rhizome (underground stem) in a cup and steep with hot water. Because of its "spiciness" you normally add other ingredients such as honey or a slice of lemon or orange.

Lemon balm (*Melissa officinalis*)

Lemon balm is a bushy, hardy, herbaceous perennial to 3 ft that tolerates shade. The small white flowers attract bees. Use the leaves as a flavoring to fish and poultry dishes, or add in salads. Herbal tea made from an infusion of the leaves is known to reduce stress and have a calming effect.

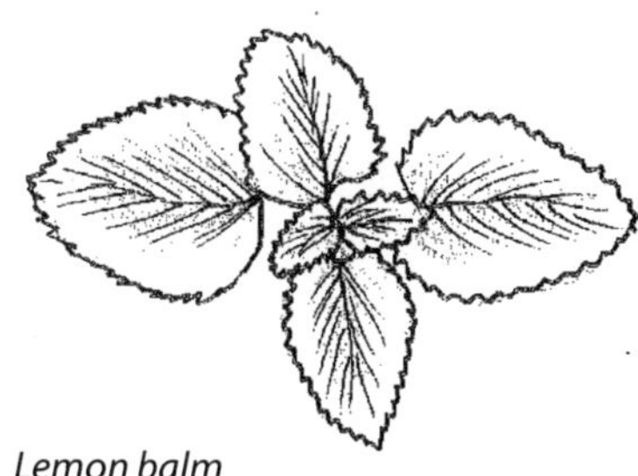
Lemon balm

Lemongrass (*Cymbopogon citratus*)

Although common in tropical regions, lemongrass can be grown in other climatic zones provided it receives adequate water and shelter, especially from frosts.

There are quite a few species of *Cymbopogon* and some are used for the production of citronella oil (insect repellent) while others are used for cooking and making drinks. The common lemongrass (*Cymbopogon citratus*) is a perennial grass that has dense clumps of long, tapered leaves. The leaves are used in teas and cooking and the bulb at the base of the leaves is used in Asian cooking. Plant where it can be well watered.

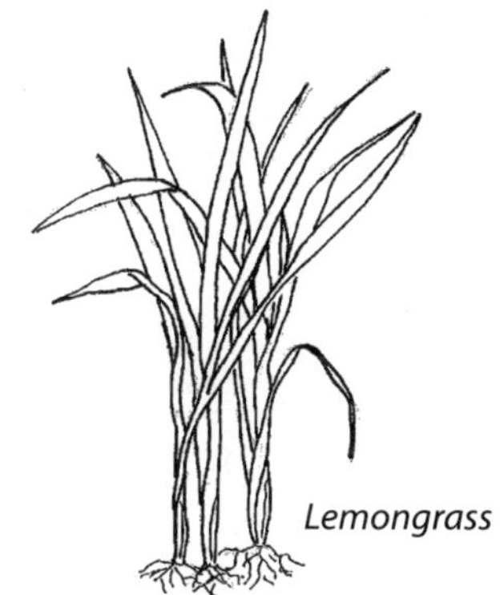
Lemongrass

» DID YOU KNOW?

The "lemon" flavor and smell of many herbal teas is due to a range of organic substances. These include citral and citronellal. Rose-scented bushes often contain geraniol or nerol, while linalool gives the sweet smell found in lavender.

Lemon verbena (*Aloysia citriodora*) [syn. *A. triphylla*]

A native from South America, lemon verbena is an evergreen shrub in warm, humid climates but can be deciduous in colder climates. The narrow, light green leaves give a lemon fragrance to any drink or dish. Dried leaves can also be used in sachets and sleep pillows.

After flowering each year, trim the shrub to maintain bushiness because it can become scraggly.

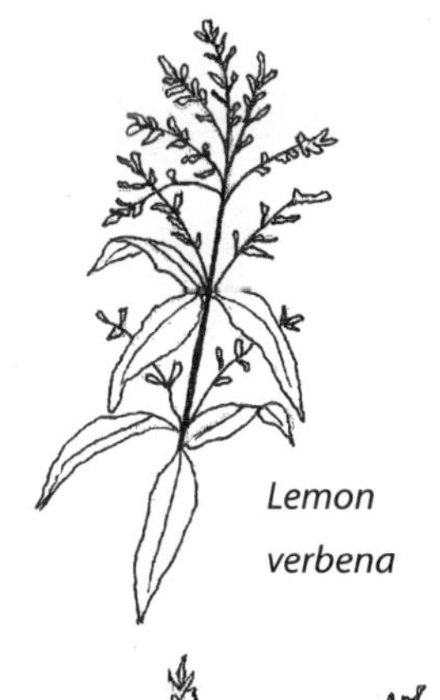
Lemon verbena

Mint (*Mentha* spp.)

Most mint plants can be used for herbal teas, except some of the perfume varieties like eau de cologne.

My favorites are peppermint (*Mentha × piperata*) and spearmint (*Mentha spicata*). Both have strong flavors and make a

Spearmint

refreshing tea, either hot or cold. Keep the strained tea in a refrigerator or make into ice blocks.

All mint varieties need moist environments and should be contained in hanging baskets or pots, unless you want them to spread. Mints are also used to make garnishes and sauces.

Thyme (*Thymus* spp.)

Common thyme (*Thymus vulgaris*) is a low evergreen shrub and some varieties are ground covers. Many hybrids and cultivars exist, and most prefer sunny locations in well-drained soils. Leaves or sprigs are used in cooking, and oil is extracted from the leaves and is a known antiseptic. A tea, made from infusing the herb in water, is a refreshing drink and a useful mouthwash.

Lemon thyme (*Thymus citriodorus*) also has many cultivars including variegated lemon, orange and lime — all of which make great scented herbal teas.

There are many "lemon" flavored teas and if you want to try more, the Australian rainforest tree, the lemon-scented myrtle, *Backhousia citriodora,* is worthy.

Many scented geraniums also make good herbal teas. You will just have to experiment.

Pest-repellent herbs

So many herbs contain oils and other chemicals that repel pests. Pest control is not only about repelling pests, it is also about attracting predators into the garden. Here are just a few common herbs that do one or the other, or both.

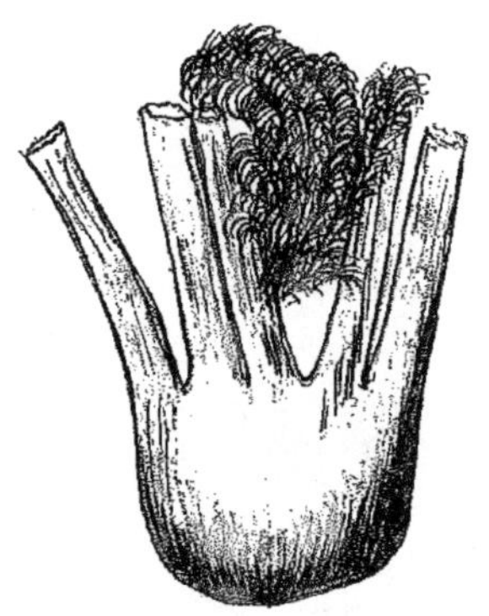

Fennel bulb

Fennel (*Foeniculum vulgare*)

Fennel is one of the oldest cultivated herbs and it attracts a range of predators as well as repelling fleas and other pests. However, it is better known for its culinary uses (the leaves, bulb and seeds are eaten).

Garlic (*Allium sativum*)

When planted around a tree or food crop, garlic is said to repel borers, mites and other sap suckers.

Rich in organic compounds that give it its characteristic odor, garlic (along with most other alliums like onions and leeks) has traditionally been used for pest control, as well as having medicinal and culinary uses.

Lavender

Lavender (*Lavandula* spp.)

Common lavender (*Lavandula angustifolia*) and lavender oil are very effective insect repellents.

Lavender leaves and oil keep moths away from clothes, weevils from cupboards and fleas from pets and carpets. Common lavender is an attractive perennial shrub with mauve (or occasionally white) flowers.

Lavender is a hardy perennial that is reasonably drought tolerant. There are many varieties and one of my favorites is green lavender (*L. viridis*), which also has a yellow variety.

» DID YOU KNOW?

Some flower shapes attract predatory and parasitic insects. Particular flowers may be shallow and contain both nectar and pollen, so predatory insects, such as hoverflies, wasps and ladybugs, can obtain immediate energy while they hunt for bugs to eat. Other plants, such as fennel, coriander and dill, have flower clusters called umbels (that resemble upside-down umbrellas), which are open nectaries offering insects easier access to nectar.

Nasturtium

Nasturtium (*Tropaeolum majus*)

Nasturtiums are rambling ground cover plants. Their yellow, red and orange flowers and green leaves are reputed to repel aphids and red spider mites from nearby plants. All nasturtiums have mustard scented foliage and both the leaves and flowers are edible. We can eat the seeds as a substitute for capers.

Pyrethrum

Pyrethrum (*Chrysanthemum cinerariifolium*) [syn. *Tanacetum cinerariifolium*]

Pyrethrum is a small, hardy, perennial bush widely used in organic pest control products.

To make an insect spray, cover chopped flowers with mineral oil and soak overnight in a dark place. Strain and store in a cool dark place. To make up the spray, add to the concentrate six parts of water and a few drops of sesame, or another similar, oil.

Pyrethrum spray has low toxicity and is biodegradable. It doesn't keep well so use it all within a few days of making.

Rue

Rue (*Ruta graveolens*)

While rue can be used in culinary food dishes and to treat a host of medical conditions, it is also used for pest control. Rue is reputed to repel fleas, flies and beetles, and it attracts predatory insects into the garden.

Santolina

Santolina or Cotton lavender (*Santolina chamaecyparissus,* or *S. incana*)

Santolina is an attractive, compact, small shrub with gray foliage and yellow button-like flowers. It prefers well-drained soil, tolerates full sun and is drought hardy. The leaves are used to repel moths (in clothing drawers and pantry shelves). Both the leaves (fresh or dried) and sprigs are suitable.

Scented pelargonium

Scented geraniums (*Pelargonium* spp.)

There is a wide variety of scented geraniums available. They are all good, general pest repellents as pests are deterred by the strong smell of their leaves. Some scented geraniums can also be used in herbal teas and to add flavor in cooking.

Scented geraniums grow best in well-drained, fertile soil in full sun but can withstand dry conditions quite well.

The leaves from various lemon-scented pelargoniums (containing citronella oil) can be rubbed on the skin to deter mosquitoes. Scented geraniums are often confused with true geraniums (or cranesbill), but while the two types of plants are related, they are distinctly different.

Tansy (*Tanacetum vulgare*)

Tansy is a small, perennial herb that can be invasive as it has spreading underground rhizomes. Although tansy has both culinary and medical uses, it is primarily used as a repellent for flies, fleas and ants. Tansy oil contains several aromatic substances that repel pests and has fungicidal and antibacterial properties. The leaves are rich in nutrients and can be used as a catalyst in compost making.

Tansy

Wormwood (*Artemisia* spp.)

True wormwood (*Artemisia absinthium*), a shrub to 6 ft, has been used for centuries as a pest repellent. Sprigs can be left in the nesting boxes of chickens (to repel lice) or wormwood "teas" can be sprayed on vegetables to repel snails, moths and other pests. It is often ingested by birds and animals to deworm intestinal tracts, but is toxic to humans if consumed.

Roman wormwood (*Artemisia pontica*), sea wormwood (*A. martima*) and southernwood (*A. abrotanum*) can all be used for pest control but are less potent.

Many other herbs repel insects and other pests and these include balm of Gilead, elderberry, catnip, dogbane, mint, marigold, rosemary and basil.

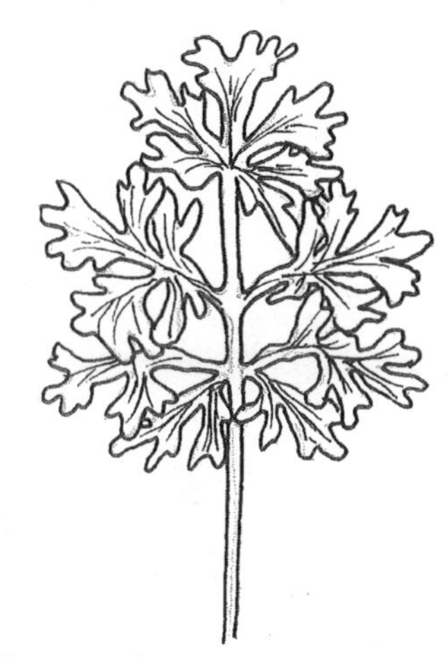

Wormwood

A herb medical cabinet

Consult any herb book or website and you will find an exhaustive list of the greatest ever medicinal herbs. Unfortunately, many are not readily available from your local nursery, or allowed in some countries or shipped between some states. Some are even declared as noxious weeds. So herbs like ginseng, burdock, wild quinine, St John's wort, mullein and great yellow gentian are not discussed here.

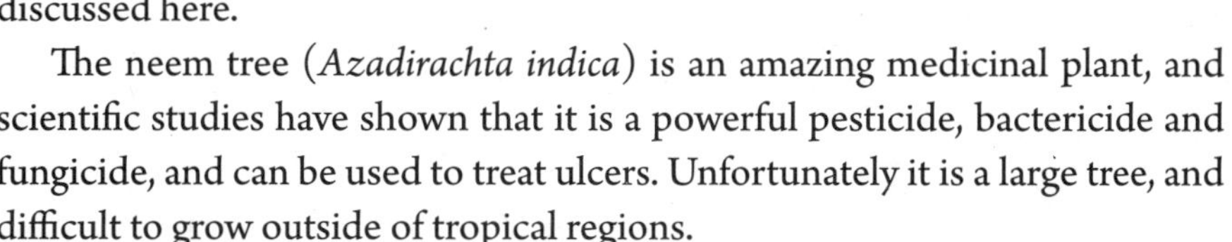

The neem tree (*Azadirachta indica*) is an amazing medicinal plant, and scientific studies have shown that it is a powerful pesticide, bactericide and fungicide, and can be used to treat ulcers. Unfortunately it is a large tree, and difficult to grow outside of tropical regions.

So we need to find universal plants that survive in a wide range of soils and climates, or plants small enough to be looked after in warm and sheltered places.

Some plants such as aloe vera have folklore medicinal uses but scientific studies have not fully supported the "magical" gel that is commonly used in drinks, creams and lotions.

While I can attest to successfully using the gel on burns and cuts, there doesn't appear to be much evidence supporting its healing powers. The same can be said for comfrey, ginkgo, calendula, sage and vervain and the vast majority of herbs for that matter.

There has been considerable research into the different types of organic substances found in herbs, possibly due to researchers trying to discover new compounds or the ultimate cure for AIDS or cancer. Not much research, surprisingly, has been done on the herb itself for its action to treat diseases and conditions.

The following herbs are common enough for the home garden and some of these are also edible. If you wish to try and use these plants you should consult other sources to see how they are best prepared.

For some plants, the flowers or fruits contain active ingredients while for other plants the leaves or both the flowers and leaves are utilized.

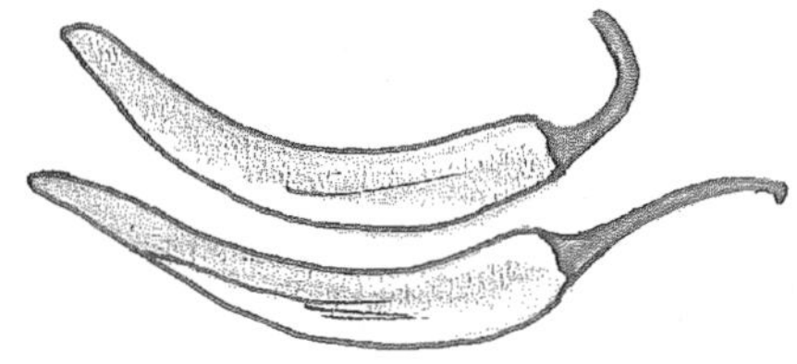

Cayenne pepper

Cayenne Pepper (*Capsicum annum*)

Cayenne pepper is a variety of the common chili pepper and bell pepper. It is a "hot" variety that is typically red in color, and contains the active ingredient capsaicin. Capsaicin has been shown to reduce pain, especially for those suffering from shingles or recovering after operations.

The powdered form of various types of peppers is called paprika.

Feverfew

Feverfew (*Tanacetum parthenium*)

There is good evidence that feverfew relieves migraines and headaches, although it doesn't seem to relieve fever as the name suggests. A couple of leaves are normally eaten, but these can be bitter, so it is best to consume them with some other food. Feverfew is believed to be a preventative treatment, so eating these daily is a worthwhile strategy for those who suffer frequent migraines. Some people simply sniff the flowers and leaves, as feverfew has a very distinct odor.

» DID YOU KNOW?

Whenever scientific and medical studies are undertaken it is essential that the participants (subjects) do not know whether they are given the drug or herbal extract or whether they have the placebo (some harmless substance). All those that are given the placebo believe they are taking the drug. This is called a blind or blinded experiment. When the researchers are also unaware of what medicine is the drug and what is harmless, this is called a double blinded experiment. This is normal practice in medical research and these types of clinical trials establish if a particular herbal medicine has any meaningful results.

Peppermint (*Mentha* x *piperita*)

Peppermint oil has been shown to improve digestive disorders, such as irritable bowel syndrome, and spasms in the intestines and stomach. Peppermint leaves can be used to make a tea but the oil is usually extracted by steam distillation of the leaves. The oil contains many compounds with menthol being the main ingredient. Peppermint oil has also exhibited antibacterial and antiviral properties and has been used to treat bad breath and dandruff.

Peppermint

Tea tree (*Melaleuca alternifolia*)

Tea tree is an Australian shrub or small tree that can be pruned to maintain a compact size. The oil from the leaves has been shown to be reasonably effective against fungal infections, such as athlete's foot. It is an effective antibacterial agent and a useful disinfectant.

There is some evidence that it is also effective in the treatment of other skin infections such as acne, warts and vaginal infections. Tea tree oil has been shown to alleviate inflammation and itching from dandruff.

Tea tree

» DID YOU KNOW?

The leaves of many Australian native plants contain oils with proven antiseptic (kills germs), antifungal and anti-inflammatory properties. These plants include species of melaleuca, eucalyptus, callistemon and kunzea.

Yarrow

Yarrow (*Achillea millefolium*)

Yarrow is a small, perennial plant with white flowers and a vigorous, spreading root system. Common garden cultivars can also have yellow or red flowers.

Yarrow leaves grow from underground rhizomes and are feathery and aromatic.

Yarrow oil has been shown to have antibacterial and antifungal properties and there is some evidence that yarrow extracts help with wound healing.

As more research is undertaken on the numerous beneficial properties of common herbs, by exploring their traditional uses in many cultures throughout the world, more herbs and their products can be added to the home medical cabinet.

OTHER USEFUL PLANT GROUPS

8

WHILE A FEW INDIVIDUAL PLANT SPECIES are listed here, sometimes plant groups (genus or family) are discussed, primarily because several species in that group are very useful plants. While the neem tree and drumstick tree are specific species, the genus Yucca and the tribe Bambuseae (bamboo), for example, are discussed.

» DID YOU KNOW?

When organisms are classified they are placed in a taxonomic rank. When I was younger I learned that Kings Play Chess On Flat Glass Sheets (kingdom, phylum, class, order, family, genus, species), but as more research has been undertaken on organisms and their relationships, other categories of groups have been created. As an example, starting with family, we now include tribe, genus, section, series, species, variety, form (and even subfamily, subtribe, subgenus, subseries and subspecies and so on).

Dryland plants

Aloe (*Aloe* spp.)

Aloes are succulents and store water in their fleshy leaves. Most have reduced stems, with a rosette (circular ring) of leaves emerging from the ground.

The gel from the leaf of the most well-known aloe, aloe vera, is used to soothe insect bites, burns and cuts, and ongoing research is investigating the validity of these traditional medical uses.

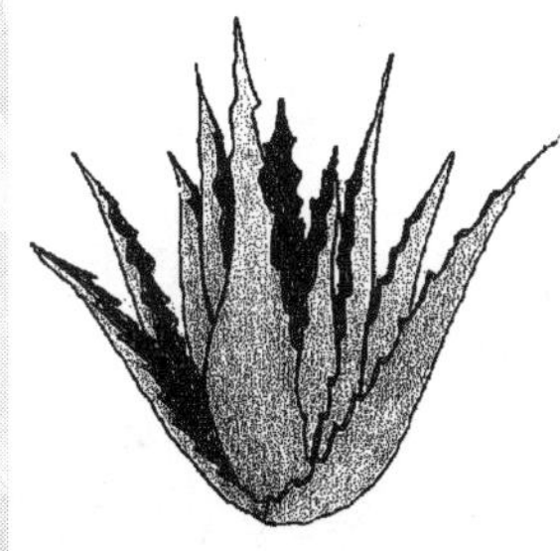

Aloe vera

Cacti (Cactacerae family)

Cacti include a wide range of species that are found in deserts throughout the world. These succulents are a source of water, food, medicine and useful tools (such as spines for fish hooks, needles).

Cacti have spines (modified leaves), which protect them from grazing animals. The stems contain chlorophyll for photosynthesis and they are water storage organs.

Date palm

Date palm (*Phoenix dactylifera*)

A truly multifunctional tree of the desert. Besides the well-known edible fruits of the date palm (which are also used in the production of syrup, breads and cakes), the leaves of the tree can be used to make furniture, thatching for houses and baskets, and the sweet sap of the tree can be tapped to make palm sugar, molasses and wine.

Date palm seeds can be ground into flour, the leaf sheaths can be made into rope and even the flowers are edible.

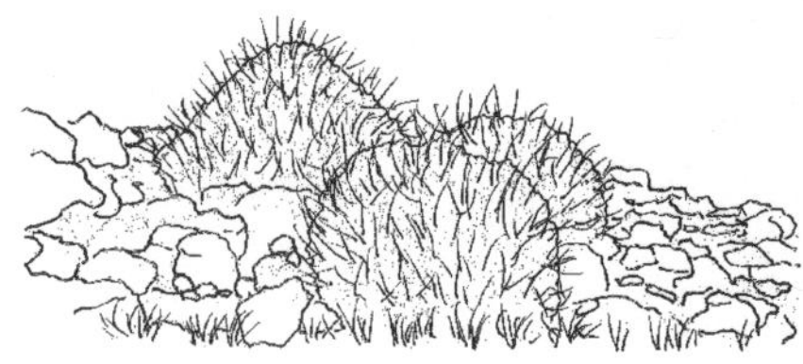

Triodia *spinifex in central Australia*

Spinifex (*Spinifex and Triodia* spp.)

A number of species of tough grasses that cover about one-quarter of Australia, spinifex thrive in the poorest of soils and the driest of conditions. They can be found in deserts (*Triodia*) and sand dunes (*Spinifex*). Their extensive roots hold soil, their seeds are food for many birds and animals, and their spiky nature provides habitat for lizards, birds and small mammals.

Aborigines extracted resin from the base of *Triodia* stems and melted it to make an adhesive (gluing stone chips onto wooden spears) and they ground the seeds into flour.

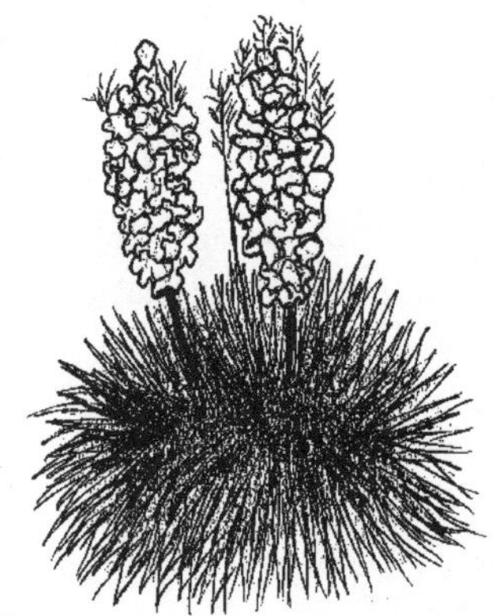

Yucca

Yucca (*Yucca* spp.)

Yuccas are from the Americas, and different species exist in a wide range of climates and soils.

While these plants are mostly grown today as ornamentals, Native Americans made soap using the detergent-like compound from the sap of the yucca roots.

Yucca stalks, blossoms and seeds were also eaten. The fiber in the leaves is used to make sandals and simple rope, and dried leaves are good to help start a fire.

Cool climate plants

Alder (*Alnus* spp.)

Fast-growing, short-lived deciduous trees that, although not legumes, are nitrogen-fixing and create dark soil. Species range in height from 15–80 ft.

Alders are used as a nurse crop for other trees and they provide shelter, mulch and compost material. Wood is used as firewood, and the timber burned to "smoke" fish and other seafoods. Alder bark contains anti-inflammatory compounds and the hard wood is used to make furniture and musical instruments.

Amaranth (*Amaranthus* spp.)

Upright annuals up to 3 ft of which grain amaranth (e.g. *A. hypochondriacus*) and leaf amaranth (e.g. *A. cruentus*) are the most valuable. Grown in full sun or even partial shade, amaranth are fast-growing shrubs.

Grain amaranth is a gluten-free, high-protein crop (14%) rich in calcium, magnesium, phosphorus, manganese and iron, and vitamins folate, riboflavin and B6. The seeds are eaten popped or ground up into flour.

Leaf amaranth is eaten raw or cooked, and the tasty red and green leaves are high in vitamins A, C and K, and minerals calcium, potassium and manganese.

A dye can be made from the red flowers, some species have edible roots and many species are used as chicken and stock fodder. Leafy varieties can be turned into silage and these species are usually good as cover crops.

Amaranth

» DID YOU KNOW?

Amaranth is known as a grain, but it is technically a seed. Grain crops are derived from grasses (cereals), but many species of broadleaf plants have seeds that are used in the same way as the cereals. The seeds can be eaten, ground into flour and cooked. Amaranth, along with buckwheat, quinoa and chia, are collectively known as pseudocereals (false cereals).

Black locust (*Robinia pseudoacacia*)

Fast-growing, long-lived deciduous tree 30–60 ft that is very hardy and survives in poor soils. Nitrogen-fixing, but tends to sucker and can become weedy, has thorns.

Robinia

Very hard, heavy wood, resistant to rot (good for fence posts), excellent firewood (burns hot and slow), bee forage (honey production) and seed for poultry.

Robinia trees can also be coppiced and the leaves fed to farm animals, while the flowers can be steeped in a cup of water to make a tea or battered, fried and eaten. The roots can also send up suckers, eventually forming an impenetrable thorny thicket. One approach is to graze stock that eat the young suckers.

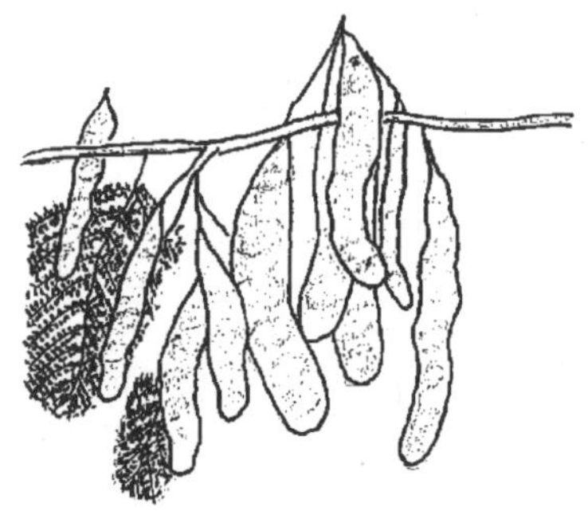

Cape wattle

Cape wattle (*Paraserianthes lophantha*) (formerly *Albizia lophantha*)

Fast-growing, nitrogen-fixing, short-lived (68 years) small tree (12–20 ft). Leaves have 20% protein, and while they are poorly digested, they still make a good supplement for other fodder.

Attracts ladybugs for aphid control, extract soap from leaves, good firewood.

Also useful as a windbreak, shelter tree and soil conditioner. Great understory plant.

Flax (*Linum usitatissimum*)

Flax

Flax is an extremely useful annual plant that grows to 3 ft. The fibers of the stalks are high in cellulose and stronger than cotton. Flax seeds are very high in protein, polyunsaturated fats, fiber, magnesium, phosphorus, iron, zinc and thiamine.

Linseed oil is extracted from its seeds, and while it is an edible oil, it is mainly used as a sealer for timber and to make putty and varnishes, and some years ago the floor covering "linoleum." Beware: as the oil quickly oxidizes in air and several chemicals are usually added to it in its processing, it is not that suitable for human consumption.

Flax fibers make linen, are blended with wool and synthetic materials to make a range of fabrics, and can be used to make canvas and paper.

Hawthorns (*Crataegus* spp.)

About 200 species of shrubs and trees found throughout the Northern Hemisphere. Long-lived hedge and windbreak, habitat and shelter for wildlife, tough but thorny and tends to sucker, and tolerates shade and poor soils. Trees can be coppiced for continual supply of firewood, fencing timber and tool handles.

Edible berries that are made into jam, wine, jellies or juiced for drinks. Some species have edible leaves that are added to salads when young and tender.

Hawthorn

Hemp (*Cannabis* spp.)

There are many cultivars of hemp that contain little or no psychoactive drug compounds, and are more than suitable to make clothes, rope and paper, biofuel and oil from its edible seeds (seeds mainly for animal and bird feed).

Hemp plants are fast growing and mature after three to four months. They require less fertilizer, weed control, insecticides and herbicides than some other crops such as cotton.

Hemp

Hickory (*Carya* spp.)

Large deciduous nut trees from Asia and North America that typically produce tough and hard timber, which is used for tool handles, bows, skis and wheel spokes. High energy wood for burning in stoves and for smoking meats. Not all species have palatable nuts for humans and many are only used as an adjunct to animal feed.

The pecan (*C. illinoinensis*) is the most important commercial tree of this genus. Pecan nuts are high in fiber, protein and fat (mainly mono- and polyunsaturated), and they contain reasonable amounts of magnesium, phosphorus, copper, manganese, zinc and thiamine.

Paulownia or powton (*Paulownia* spp.)

There is debate about how many species of paulownia exist (estimates of 6–20) as there are large numbers of crosses and variants as paulownia trees have been cultivated for a few thousand years.

All are quick-growing, drought-resistant deciduous trees, with some growing up to 50 ft. Characteristic very large leaves that are useful fodder for stock, light but durable timber used to make fine furniture and chests, and the trees are used in agroforestry to shelter various crops.

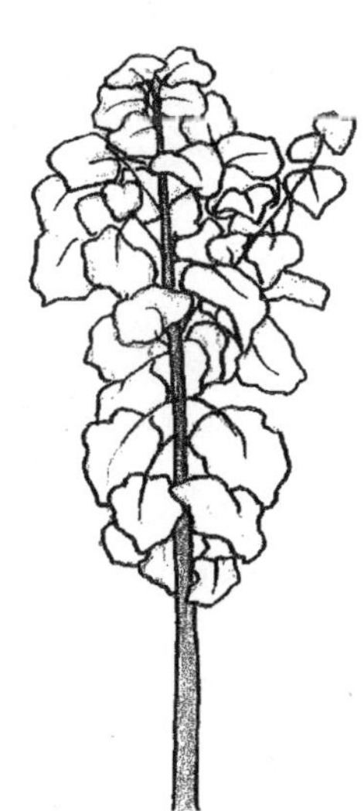

Paulownia

Caution: not a good tree beside buildings or ponds as the root system is highly invasive and capable of lifting concrete. New shoots erupt from damaged roots.

» DID YOU KNOW?

The paulownia group of trees are the fastest growing in the world. They can grow up to 13 ft in one year.

However, the fastest growing plants are the bamboos. Larger varieties can grow nearly 3 ft in one day (that's half an inch every 20 minutes), and as the 100–130 ft high culms (stalks) reach maturity in one year, the average growth rate over the whole year is 4 in a day.

White cedar or Cape lilac (*Melia azedarach*)

A relatively small (to 30 ft), short-lived (20 years) deciduous tree, white cedar is related to the neem (same family) and has similar insecticidal properties (but in milder levels).

Cape lilac

The timber is resistant to termite attack and is used to make furniture, it coppices well (useful firewood source), and the berries (poisonous if eaten, except for some birds) and leaves can be used to make a garden spray for pest control. However, it spreads easily and can become invasive.

Many other useful plants are grown in the temperate and milder regions of the world and these include goji berries, blueberries and blackwood (*Acacia melanoxylon*).

Warm humid climate plants

Amla (*Phyllanthus emblica*)

Amla, also known as Indian gooseberry, is a medium-sized deciduous tree. While it can grow in a range from light to heavy soils, it does need protection from excessive sunlight and strong wind.

Dried amla berries are soaked in oil and the extracted oil-soluble vitamins are used in shampoo and hair conditioner. Its large fruit, although tart, can be picked and eaten, contains high levels of antioxidants and therefore has health benefits. The timber and branches are used as firewood.

Bamboo

There are over 1,000 species of bamboo, which are generally grouped into two types — clumping and running. Generally, the tropical and subtropical varieties are clumpers and the temperate varieties are runners.

Running bamboos spread and can invade under the fence into neighboring yards, so it is best to use them in larger properties or where stock can eat and maintain them to a confined area.

Bamboos belong to the grass family, but unlike other grasses that have their leaf sheath wrapped around the stem, bamboos have a short petiole that attaches the leaf to the branch.

You can further identify running types as most have a flattened edge on alternate sections of the culm (the upright stalks or stems) while clumpers tend to only have round culms.

Bamboo is a true multifunctional plant. Some are used as food or fodder, others as shelter and windbreak, and many have structural uses to make buildings, bridges, pergolas and as reinforcing in concrete.

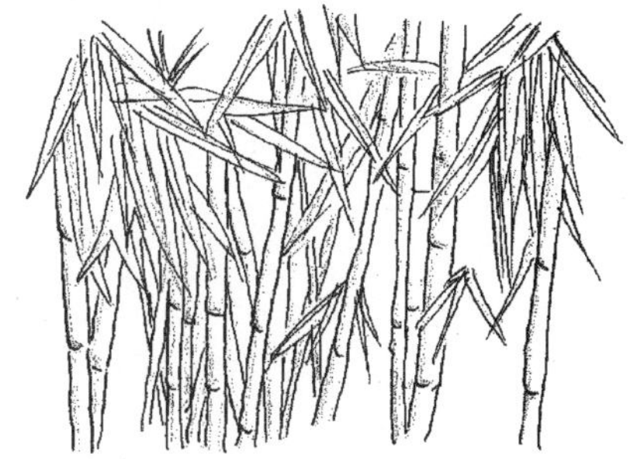

Bamboo

Cacao (*Theobroma cacao*)

Cacao (pronounced "kah kow") is also known as the cocoa (pronounced "ko ko") tree and it supplies cocoa, chocolate and cocoa butter. An evergreen, understory tree in rainforest areas, the cacao produces large pods that contain many seeds or "beans."

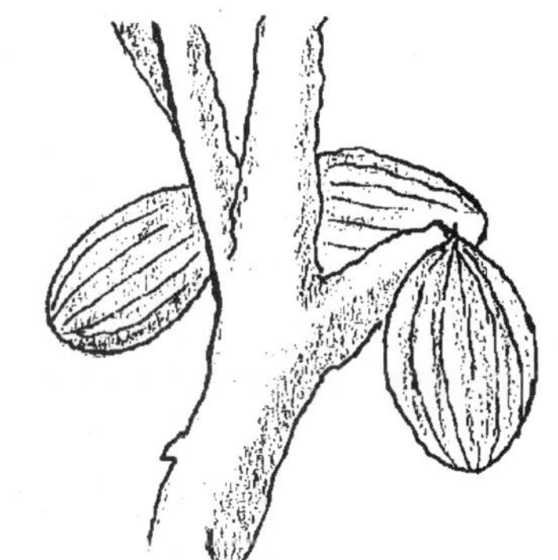

Cacao pods

The seeds contain about 40% fat as cocoa butter and this is the main ingredient of chocolate. The sweet, white pulp surrounding the seeds is used to make juices and jellies.

Cacao trees are normally established under shade and they are very susceptible to both cold weather and high temperatures. They are very hard to keep alive outside of tropical areas, but who wouldn't give them a go if they can make chocolate?

Cassava (*Manihot esculenta*)

The tuberous roots of this plant are boiled like potatoes and eaten for their high starch content. The dried starch (called

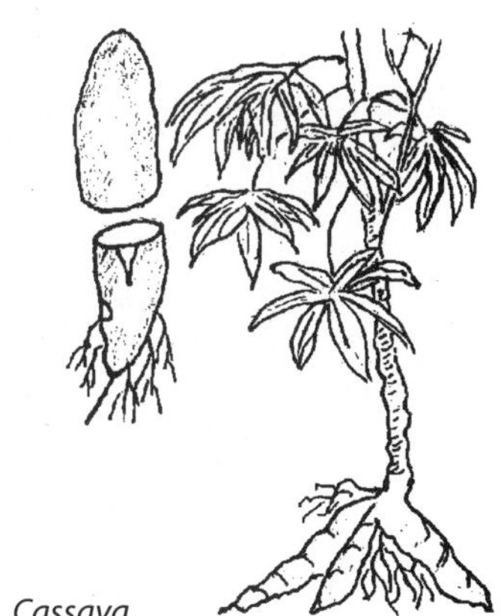

Cassava

tapioca) is also used as a thickener. Cassava must be cooked to detoxify it as it does contain cyanide.

Cassava plants are typically grown as an annual and the whole plant pulled to harvest the roots. The stems can be sliced into cuttings and then planted for next year's crop. This short woody shrub can tolerate a wide range of soils, can withstand drought and survive in low rainfall areas.

The tubers also contain fair levels of vitamin C and calcium.

» DID YOU KNOW?

Underground food crops can be either bulbs, corms, roots, tubers or rhizomes, all of which can act as storage organs for the plant. Onions are bulbs as these are layers of modified leaves, taro is a corm as this is a swollen stem at just below ground level and is a strategy to survive drought or adverse seasonal weather (the plant may die back in winter and will regrow in warmer conditions).

While a carrot is a swollen taproot, the cassava and sweet potato have swollen tuberous roots. These swellings are enlargements of secondary or lateral roots, away from the main taproot. The common potato is a tuber, which is a swollen stem part, as it contains the normal parts of stems such as nodes (eyes of potato) and both have similar internal cell structure. Ginger, turmeric, bamboo and arrowroot have rhizomes, which are swollen horizontal underground stems.

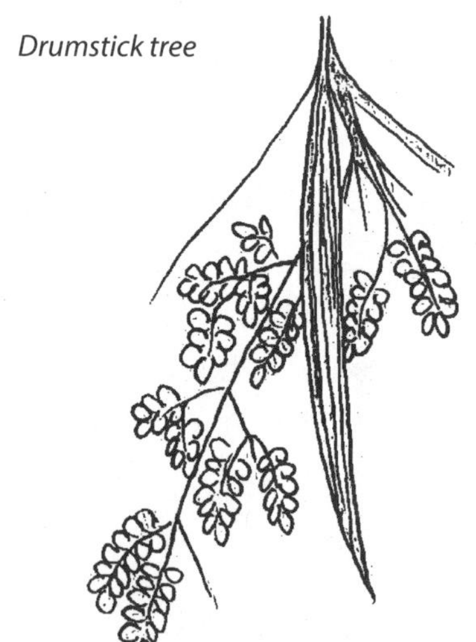

Drumstick tree

Drumstick tree (*Moringa oleifera*)

Moringa oleifera is a fast-growing deciduous tree that has edible leaves and seed pods. Fuel and cosmetic oil is derived from its seeds and some of the substances left in the seeds after the oil is extracted show promise for the disinfection of contaminated water.

Called the drumstick tree after the long slender seed pods, it is also known as the horseradish tree after the taste of its roots (which obviously taste like horseradish). It is fairly drought tolerant but doesn't survive continuous cold weather and frost.

Moringa leaves contain high levels of protein, vitamins A and C, and minerals calcium, magnesium, iron and potassium.

Ice-cream bean (*Inga edulis*)

Ice-cream bean

Fast-growing, nitrogen-fixing, evergreen tree to height of 30 ft and spread of 12 ft. Drought and frost sensitive when young. Seed pods up to 20 in or more depending on the species or cultivar. The white pith, with a cotton wool texture, around each seed can be eaten and tastes like vanilla ice cream. Wood can be used for furniture and the leaves for feeding stock.

Indian tulip tree (*Thespesia populnea*)

The Indian tulip tree or Portia tree is a fast-growing, evergreen tree (but can shed its leaves in some areas) to about 30 ft high. It does tolerate a wide range of soils and prefers coastal regions. The bark and leaves have traditional medicinal uses, and the fruits, flowers and young leaves are all edible.

The tough, fibrous bark is made into rope. Yellow and red dyes are obtained from its fruits, flowers and bark. Its timber is used to make paper and its colored wood is used to make furniture, musical instruments and carvings (including bowls and plates).

Neem tree (*Azadirachta indica*)

Neem

Neem is a fast-growing, evergreen tree that is drought resistant, produces good shade and can tolerate a fair range of soils, although it seems to prefer sandy soils. Neem trees won't survive in heavy waterlogged soils nor cold climate regions in the world.

The fruits and seeds of the neem are the source of neem oil, which is used as a lubricant, to make soap and cosmetics, and to repel insect pests in orchards. Neem leaves can be dried and placed in cupboards or pantries to prevent insect attack of clothes, seeds and food. The tender shoots and flowers can be eaten as a vegetable, the slender twigs are used as a toothbrush, the gum is a rich source of protein and used as a bulking agent in prepared foods, the wood is used to make furniture, the bark has a high tannin content and is used in the furniture industry, and its fibers can be woven into ropes. No wonder some people call the neem the most useful tree in the world.

» DID YOU KNOW?

Many seeds of tropical and subtropical plants have short viability. They are known as recalcitrant seeds and must be sown within a few weeks of collection. They do not keep well, as they dry out quickly or suffer with exposure to both low and high daytime temperatures. It is common to see some seeds germinating in mature seed pods. Recalcitrant plants include the neem, ice-cream bean and drumstick tree, as well as mango and avocado, discussed in an earlier chapter.

Rubber tree (*Hevea brasiliensis*)

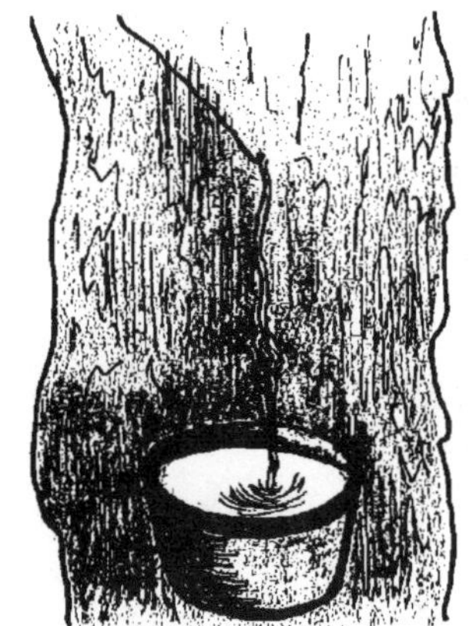

Rubber

Now grown throughout the tropical world, the rubber tree (also known as the Pará rubber tree) produces a milky white sap called latex, which is used to make natural rubber.

Special vessels in the bark spiral anticlockwise upwards and carry the latex. By carefully cutting the bark on an angle (clockwise) the vessels are intersected and latex can flow downwards and then into a small bucket.

Many other plants also contain the milky elastic sap but none has the production of this tree.

Tamarind (*Tamarindus indica*)

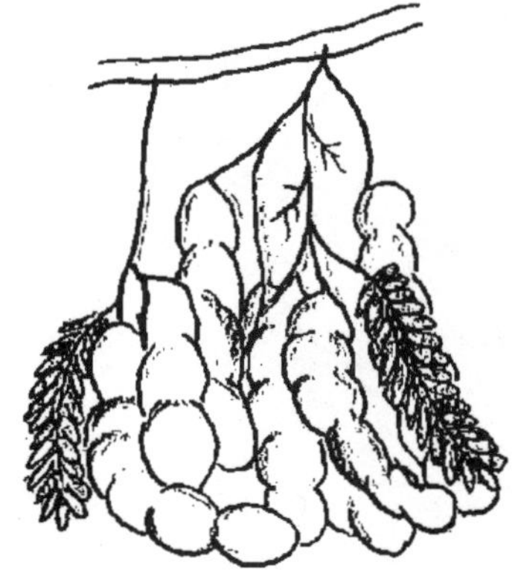

Tamarind

The tamarind is a medium-sized tree that tolerates a wide range of soils, and withstands occasional drought and salty air in coastal regions. It makes a good windbreak, but does not tolerate waterlogging, frost or shady conditions. Besides the edible pulp in the pod-like fruit, the tamarind has dark red heartwood timber that is used in many furniture applications.

The pod pulp is high in iron and magnesium with good levels of the vitamin B group, calcium, potassium and phosphorus. The fruit is used as a flavoring for many dishes, including desserts, and is made into jam and sauces. Both ground seeds and leaves are high in protein and used as fodder for animals.

Many other tropical and subtropical plants are extremely useful, including ginger, cotton, coconut, jute, sapodilla and coffee.

FODDER AND FORAGE SHRUBS AND TREES FOR FARM ANIMALS

9

MANY TYPES OF PLANTS are used to feed animals. Grain crops (oats, sorghum, millet, corn) and a whole host of grasses are the most commonly used fodder worldwide. This section mainly deals with the larger shrubs and trees.

In the US and Canada and most other countries there are several climatic zones, typically based on rainfall. So there could be areas classified as dry temperate, wet temperate, subtropical, monsoonal and alpine, for example, and as different plants tend to grow in different areas, it is not possible to always obtain, grow and nurture every tree and shrub mentioned here. Do some research and select the plants most suited to your soil and climatic region.

For simplicity I have divided these plants into three groups: those that survive in dry climates (arid, desert, low rainfall areas), those that prefer cold climates (temperate and Mediterranean) and those that tend to thrive in warmer climates (subtropical and tropical). Species are listed alphabetically by common name.

In North America there are 11 recognized climate zones, with unique flora in each zone, so trying to place particular plants into one of three zones is not ideal.

» DID YOU KNOW?

The terms fodder and forage are often interchangeable. However, some texts suggest that fodder is what is given to animals (cut and carried) while forage is what they source themselves. The following sections just talk about fodder as meaning both fodder and forage.

Fodder from trees and shrubs is especially useful during drier periods and the autumn feed gap when nutritious food is not always available. Fodder plants should not be seen as just providing food during shortfalls, but also as positively contributing to an animal's growth and weight gain.

Providing appropriate fodder on the farm negates the need to buy grains or food during these shortage periods and to compensate when adverse weather or other conditions minimize pasture production.

Ideally fodder plants should:

- be long-lived, perennial plants
- be reasonably drought tolerant (extensive root systems)
- be waterlogging and salt tolerant
- have the ability to be severely pruned or coppiced and then recover
- have high productivity of edible material (many tons per hectare)
- provide high quality nutrient-dense food
- be cheap and easy to establish, grow and harvest
- have fast growth rates
- be multifunctional (fire retardant, N-fixing, windbreak, useful shade, timber)
- provide leaves, pods or seeds in summer and autumn
- have high digestibility (easily broken down by the animal's digestive system so that nutrients become available)
- be highly palatable (pleasant to eat)
- contain no toxic parts
- not inhibit nearby plants and pastures (allelopathic relationships).

Unfortunately, many fodder species are short lived and do not have all of these characteristics, so it is difficult to find the perfect plants.

From a permaculture perspective we also look for plants that reduce wind and soil erosion, require minimal water use to maintain their growth, improve soil structure, increase the biodiversity of the farm and do not themselves become environmental weeds. The ways in which pasture and fodder plants are accessed by animals is mainly discussed in chapter 11.

Animals are selective eaters. They will eat saltbush and tree medic, for example, but if other herbaceous (grassy) plants are available these are preferred.

Young trees can contain too much sap, harmful chemicals and other organic substances that deter browsing, so mature trees and shrubs are more often eaten.

On the other hand, other types of young plants are often vulnerable to browsing and the whole plant can be killed. Whereas grasses can recover quickly if stock are moved on, trees and shrubs need more observation and management as greater damage can be done by overgrazing. Also, a particular forage may have low palatability in the wet season but become desirable in the dry period (or vice versa). So there are many factors and intricacies between plants and their consumers.

I also need to stress that some of the plants discussed here may be declared (and prohibited) weeds in your region. Furthermore, seed and plant material may not be available where you live.

You can substitute any of these for others that are more appropriate to your soils, climates and animal types (but do remember to check their environmental status).

Animals, like humans, cannot live on one food type alone. The plants discussed here should be seen as just one part of a varied diet.

The foods that are normally consumed are often seasonal, and many of the fodder species are useful additions or supplements to seasonal feed shortages, and as an alternative food source when adverse conditions or disasters occur.

Again, don't rely on one main fodder species but a combination of several different trees and shrubs along with pasture is more likely to provide a good balance of protein, energy, carbohydrates and good nutrition.

Mixtures of fodder plants have production increases of about 1 t/ac/yr for every 1.5 inch increase in rainfall. As grasses dry out digestibility decreases. This is because dry grass is high in cellulose and lignin (fiber) and low in carbohydrate and protein. Stock can then move onto other fodder species.

Eating fodder

Animals that eat plants (fodder, forage) are herbivores. They have specialized stomachs and digestive systems to break down cellulose and other tough plant components.

Animals that mow grass and other plants close to the ground (grasslands, pasture) are called "grazers," while those that chew on leaves, stems, fruits and woody twigs of the larger shrubs and trees are termed "browsers."

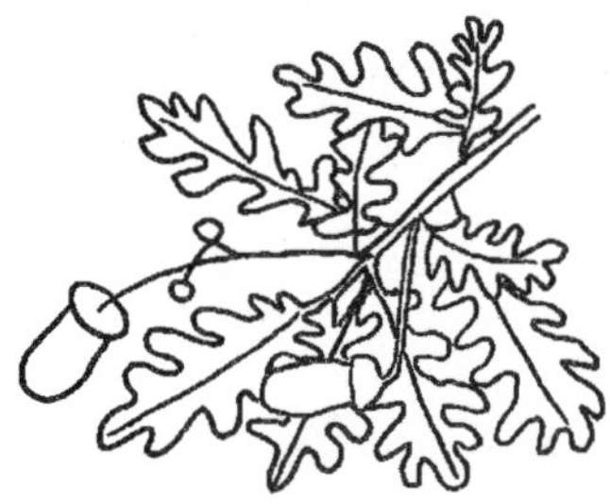

Oak produces acorns during the autumn feed gap.

Saltbush (like some other fodder species) may change its palatability from one soil type to another.

Many grazing animals can be seasonal browsers. Sheep may nibble on pasture grass most of the time, but if the grasses start to dry out and shrubs and trees are present, they will eat the leaves and branches as well.

This is typical of most animals — if their preferred food is not available they will eat something else. Most grazers will also browse and even ringbark trees if their diet lacks certain minerals.

Common farm animals that graze on low-lying plants include sheep, geese, rabbits, horses, donkeys and cattle. Farm animals that are browsers include goats and deer. Non-farm animals such as giraffes and camels also fit into this category.

Browsing animals can have an advantage over grazers when grasses die off, as they can turn their attention to woody plants.

Grazers have an advantage as there is often more nutrition in the young, rapidly growing grasses than in old mature shrubs and trees, which tend to be very woody.

Even though grazers and browsers have their own niche in the ecosystem, both groups of animals can interchange their roles to some degree.

If farmers can't provide enough food, they may need to de-stock during lean times (either sell or agist elsewhere) or store hay and silage for later use. Fodder trees and shrubs should be seen as important supplements to pasture grazing.

» DID YOU KNOW?

Sheep, goats and cows are ruminant animals, and they are foregut fermenters. They have four chambers in their stomach, which allows them to fully digest fibrous plant material. They can regurgitate their food, re-chew their "cud," then re-swallow and finally continue with the digestion process.

Horses and rabbits, on the other hand, are hindgut fermenters. They have an enlarged cecum, an offshoot of the large intestine, which permits them to digest tough plant material.

A third group of animals called pseudoruminants (alpacas, llamas and camels) have a three-chamber stomach to help them slowly ferment and digest food. They are modified ruminants as they do regurgitate and chew their cud for many hours each day.

Dryland plants

The following plants can survive in areas with an annual rainfall of less than 24 in. In some cases they are quite productive, but others only just survive. The growth habits and productivity of most plants respond to water, and to a lesser extent soil type. These same plants may be much more productive in higher rainfall areas, and possibly not so productive in clayey soils when compared to growing in sandy loams, but each species does have a preferred range of conditions in which it will grow.

Carob is a drought-hardy tree.

Name	Growth, habit	Nutritional value	Other notes, uses
Apple-ring acacia *Faidherbia albida* (formerly *Acacia*)	Deciduous legume tree (N-fixing) from arid areas of Africa, SE Asia.	40–200 lb fruit pods/tree/yr. Leaves 15% protein, 75% digestibility. Pods 11% protein, 85% digestibility.	Bee forage, useful timber, living fence (thorny).
Bladder senna *Colutea arborescens*	Low, deciduous, N-fixing shrub. Tolerates wide range of soils and conditions.	Leaves 17% protein, digestibility 60%. Highly palatable.	Recovers quickly after pruning.
Cactus *Opuntia* spp. Many species including the prickly pear group and spineless cacti groups.	Large-lobed, succulent. Spreads by seed (birds) and vegetatively.	Low protein (5%), 60% digestibility, highly palatable. Fair levels of nutrients (Mg, K, P and Ca).	Underrated as fodder. Declared species in some countries, so research to see if it is suitable for you. Thornless varieties are preferred but may revert after heavy browsing.
Carob *Ceratonia siliqua*	Slow-growing, long-lived, spreading, evergreen tree. N-fixing. Male and female plants. Pods in autumn, 40–200 lb/tree.	Pods 60% digestibility, high sugar content (5% protein, 3% fat but 70% carbohydrate). Leaves palatable.	Pods ground to make carob powder, a substitute for chocolate. Good windbreak. Fire resistant.
Combretum *Combretum aculeatum*	Small tree or large shrub.	Yield 120–400 lb/ac/yr. Leaves and fruit 15% protein, 35% digestibility.	Relatively unknown plant outside of Africa. Thin branches used in weaving.
Desert date *Balanites aegyptiacus*	Evergreen to semi-deciduous tree. Thorny. N-fixing.	Leaves (both green and dry) and fruits edible. Leaves 16% protein, 55% digestibility. Fruits 10% protein, 70% digestibility.	New shoots cooked and eaten by humans. Ripe fruits eaten and oil extracted from kernels.

Name	Growth, habit	Nutritional value	Other notes, uses
Marine couch *Sporobulus virginicus*	Perennial, low spreading grass. Highly salt and waterlogging tolerant.	Up to 10% protein. Can produce 7 t/ac/yr.	Soil stabilizer (extensive roots). Spreads via rhizomes.
Mesquite *Prosopis juliflora* Other similar *P. tamarugo* and *P. pallida*	South American legume trees introduced into Africa and Asia.	2.6 t/ac/yr. 15% protein in leaves and pods. All plant parts highly digestible.	Bee forage, bean flour used by humans, wood for furniture. Declared weed in some regions. Check with local authorities.
Saltbush *Atriplex* spp. e.g. *A. nummularia*	Spreading shrub. Most species 3–6 ft.	Some species have poor digestibility and leaves contain too much salt. *Atriplex nummularia* 20% protein, digestibility 70%, up to 2 t/ac/yr.	Used in many arid countries as fodder supplement during summer and autumn.
Wattles *Acacia* spp. e.g. *Acacia aneura*	Evergreen shrubs and small trees. N-fixing.	*A. aneura* 10% protein, moderate palatability and 50% digestibility.	*Endemic acacias* are also browse plants in Africa. Not all wattles are appropriate fodder species.

Table 9.1. Generic fodder plants for dryland areas.

Cool climate plants

Cool climates have rainfall between 24 and 40 in, and include wet and dry temperate and Mediterranean climate zones.

There are many plant species that are used as fodder for animals. Some plants are specific for particular animals and some plants are eaten by many different types of animals. Here are a few generic fodder plants.

Name	Growth, habit	Nutritional value	Other notes, uses
Ash *Fraxinus* spp.	Medium to large trees, mainly deciduous, species throughout northern hemisphere.	10–15% protein, 70% digestibility. Production from leaves 1 t/ac/yr.	Hardwood timber, firewood.
Elms *Ulmus* spp.	Fast-growing deciduous or semi-deciduous large trees.	Leaves 10% protein (seeds 45%), 50% digestibility.	Landscape trees, useful timber.
Mulberry *Morus* spp.	Fast growing, extensive root systems.	Leaves protein 15–25%. Digestibility 70–90%, highly palatable and sought-after by stock.	Good productivity in warm, humid climates too.

Name	Growth, habit	Nutritional value	Other notes, uses
NZ mirror bush *Coprosma repens*	Fast-growing evergreen shrub, dense, spreading to 10 ft high and wide (but can grow taller). Partial shade tolerant, salt and wind resistant.	Good supplement for other fodder. High protein and digestibility. Fruit and seeds for poultry, leaves for most animals. David Holmgren calls this "chocolate for animals".	The domatia (holes) on underside of leaves are refuges for predatory mites, so useful to plant in orchards as part of pest management strategy.
Oaks *Quercus* spp.	Slow-growing, deciduous large trees.	Acorns 5–10% protein. 20 lb/tree. 40–50% digestibility (contains tannins). Leaves also eaten.	Excellent timber, summer shade. Acorns drop in autumn when stock need extra carbohydrates.
Poplars *Populus* spp.	Some tolerate saline soils (Euphrates poplar). Tends to sucker. Less vigorous than willow.	White poplar *P. alba* — protein 14% digestibility 77%. Up to 4 t/ac/yr. Cottonwood 60% digestibility.	Groups include aspens, cottonwoods, balsams and subtropicals. Lightweight timber, windbreak, biomass for fuel.
Tagasaste *Chamaecytisus palmensis* (also *C. proliferus*)	Fast-growing, evergreen tree. N-fixing. Requires 20 in to be productive. Up to 4 t/ac/yr.	Leaves 15–20% protein. Highly digestible (70%) and highly palatable.	Seeds for chickens. Not tolerant of salinity, frost or waterlogging. Winter flowers for bees.
Tree medic *Medicago arborea*	Fast-growing, evergreen shrub. N-fixing.	Leaves 15% protein, 50% digestibility. Available forage 1 lb/plant.	Attracts bees and butterflies. Also suitable in semi-arid areas.
Wattles *Acacia* spp. e.g. *A. saligna*, *A. longifolia*.	Typically fast-growing pioneer, N-fixing plants.	*A. saligna* 1.5 t/ac/yr edible dry matter, with low digestibility, 15% protein.	*A. saligna* also known as *A. cyanophylla*. Is suitable in semi-arid areas.
Willows *Salix* spp.	Prefers moist habitats, tolerates heavier soils.	Up to 450 lb/tree. Leaves 17% protein.	Fast recovery after harvest.

Table 9.2. Generic fodder plants for cool areas.

Warm humid climate plants

Typically this includes plants that thrive in annual rainfalls over 36 inches in the subtropical, monsoon and tropical zones.

Some plants may not be suitable for your area. Mesquite, for example, is a declared (prohibited) plant in most states of Australia, but it is a remarkable multifunctional tree.

Name	Growth, habit	Nutritional value	Other notes, uses
Calliandra *Calliandra calothyrsus*	Small, perennial, N-fixing tree, fast growing.	Leaves and pods 22% protein. 60% digestibility. 3 t/ac/yr.	Good timber, firewood, erosion control, honey and paper production.
Centro *Centrosema pubescens*	Perennial trailing (low-lying, spreading) N-fixing herb.	20% protein, 50% digestibility. 4 t/ac/yr.	Good soil cover, complements larger forage species.
Elephant grass *Pennisetum purpureum*	Tall, clumping grass to 10 ft. Bana grass (a taller *P. purpureum hybrid*) is also excellent fodder.	10% protein, young leaves very palatable, 70% digestibility.	Spread or propagated by stem cuttings or sections of rhizomes. Useful as windbreak. Kikuyu grass (*P. clandestinum*) is a common forage pasture.
Gliricidia *G. sepium*	N-fixing, well known in tropical countries.	6 t/ac/yr. 20% protein, 75% digestibility, high palatability.	Used for firewood, living fences, shade, nursery tree.
Honey locust *Gleditsia triacanthos*	Fast-growing, deciduous tree. Thornless varieties available, but seedlings likely to develop thorns. Tends to sucker.	Pods 20–200 lb/tree. 17% protein, 60% carbohydrates (high sugar) and 7% fat.	Pod drop in autumn. Leaves are also palatable.
Indian siris *Albizia lebbeck*	Deciduous, N-fixing Asian and African legume tree that tolerates wide range of soils and climates.	2 t/ac/yr. 20% protein in leaves, 60% digestibility.	Hardwood for furniture, useful shade tree, erosion control, bee forage (honey).
Leucaena *Leucaena leucocephela*	N-fixing colonizer of disturbed ground. Can be weedy in some areas. Faster growth in higher rainfall areas.	0.5–1.5 t/ac/yr. Pods 25% protein, digestibility 80%. Leaves 20% protein, 75% digestibility.	Cattle forage in rangelands (but not as sole feed). Recovers from heavy browsing.
Pigeon pea *Cajanus cajan*	Tends to be a short-lived perennial shrub to 10 ft.	Leaves 15% protein, 60% digestibility. Seeds 20% protein, 90% digestibility.	Overgrazing will kill plants. Best to prune to feed animals. Typically grown for human food with surplus feed to stock.
Sesbania or Agati *Sesbania grandiflora*	N-fixing, moderate recovery after grazing.	20 t/ac/yr, 65% digestibility, 20% protein, highly palatable.	Smaller *S. sesban* also forage and green manure crop in low rainfall areas.
Sunn hemp *Crotalaria juncea*	10 ft shrub.	Leaves 15% protein, 60% digestibility (seeds poisonous to eat).	Green manure crop, useful fiber, fodder is dried and is not eaten green.

Table 9.3. Generic fodder plants for a warm humid climate.

Fodder species

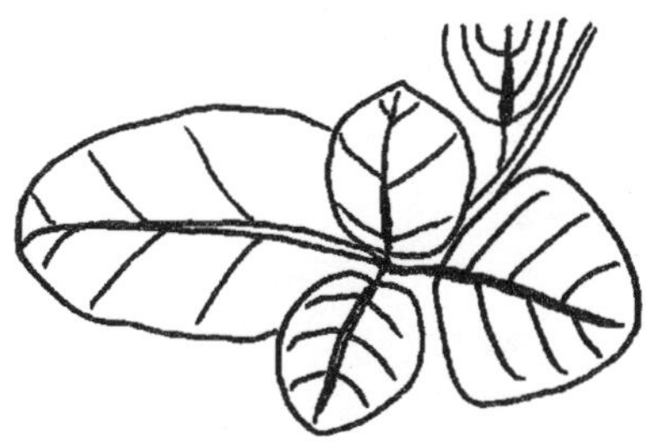

Coprosma *produces succulent leaves.*

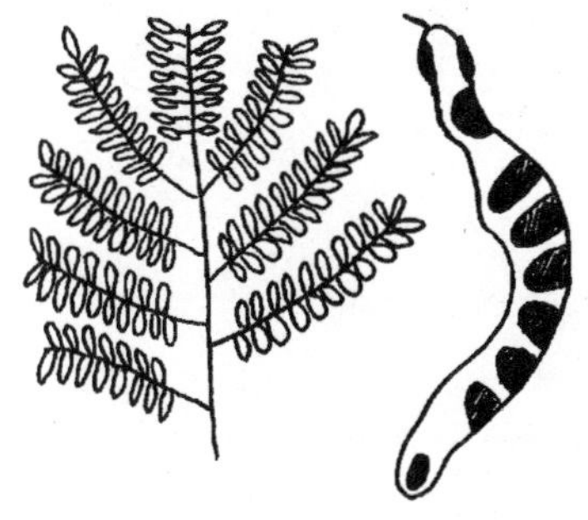

Honey locust pods contain high levels of sugars.

Acacia saligna *leaves contain 15% protein.*

Tree medic is a nitrogen-fixing shrub.

Pigeon pea provides food for both humans and animals.

Gliricidia *has high digestibility and palatability.*

ALL ABOUT WATER

10

WATER IS LIFE. We take it for granted, but every home needs to use water wisely. There are so many ways in which we can harvest water, treat it, recycle it and use it in our homes and gardens.

Saving water in the home

The message to gardeners is very simple — it's time for a fundamental change from the traditional ways of gardening.

The water wastage has to stop and there are gardeners who are treating our scarce water with little respect. We need to change our behavior and give clean freshwater its true value.

Water restrictions will continue in many places and freshwater availability will continue to fall throughout many parts of the globe in the years to come.

Let's plan and plant a water-responsible garden to help avoid a water catastrophe. Let's become smarter gardeners by examining four general areas where you could make a difference: water conservation, irrigation practices, graywater reuse and rainwater harvesting.

Water conservation

We need to learn to minimize water consumption and maximize water efficiency. Water conservation is all about ways to save us using clean town water by reducing the volume we use each day, as well as minimizing water loss through evaporation, spillage and downright neglect. Here are some handy hints.

To minimize water loss:

- Cover pools and outdoor spas.
- Use mulch in the garden — whether this is living mulch, dead plant material or other materials such as rocks or plastic sheeting.

- Add compost and organic matter to the soil to help retain water. Healthier plants are also more able to withstand drought.
- Use shady trees to protect ponds, soil and understory plants.
- Use drip line irrigation or porous (sweat) hoses to deliver water to the root zone rather than overhead sprays where wind can blow a lot of the water away.
- Add water-holding amendments to soils, such as compost and organic matter, and mineral-based amendments, such as bentonite and kaolin clays, for sandy soils.
- Water your plants either early in the morning or late in the afternoon.
- Repair leaks — a rapidly leaking tap could waste between 12–25 gal a day.

» DID YOU KNOW?

Got a leaky tap? Dripping at only one drip a second adds up to nearly 3 gal every day and if left for a year would waste about 1,000 gal: enough water to clean your teeth 14,000 times.

To minimize water demands:

- When building new garden beds, group plants that have similar water needs (hydrozoning).
- Plant drought-tolerant plants in the garden — natives and succulents, Mediterranean herbs such as lavender and rosemary, and fruit trees such as olive.
- Improve the soil. Increasing organic matter allows more water retention and more water availability to plants.
- Reduce your lawn area.
- On large garden areas it may be economical to install some water moisture sensors to activate irrigation systems. In this way, only the amount of water that plants actually need is supplied.
- Avoid washing your car on the driveway; position it on the grass instead.
- Plant windbreak trees and shrubs to reduce evaporation from the soil.
- Install a graywater reuse system, reusing wastewater from the laundry or bathroom.
- Sweep paths instead of hosing them down.

- Install permeable paving rather than solid slabs or pavers. Rainwater can move through the permeable paving into the soil.
- Take shorter showers.
- Fit water-efficient showerheads.
- Fit water-efficient aerators or flow regulators.
- Replace toilets with at least a 1.6 gal/0.8 gal dual flush — and use the half flush whenever you can. Placing a brick in the tank is an easy and economical solution.
- Use the washing machine only when it is fully loaded.
- Fit a pressure reduction valve to reduce pressure in the town water supply.
- Purchase a front-loading washing machine. These tend to use about half the water of a top-loading machine, i.e. 13 gal instead of 21–25 gal per cycle.
- Install a rainwater tank and use this water to flush toilets and provide a laundry source.

Undertake a water audit and estimate your household's total water usage. Monitor the duration of showers, estimate bath volumes, record the number of loads of washing per week, and determine how much water goes onto the garden, how much on toilet use each day and how much is used in the kitchen. Work out how much you could cut down to see how your consumption could be reduced. Aim for 50% reduction.

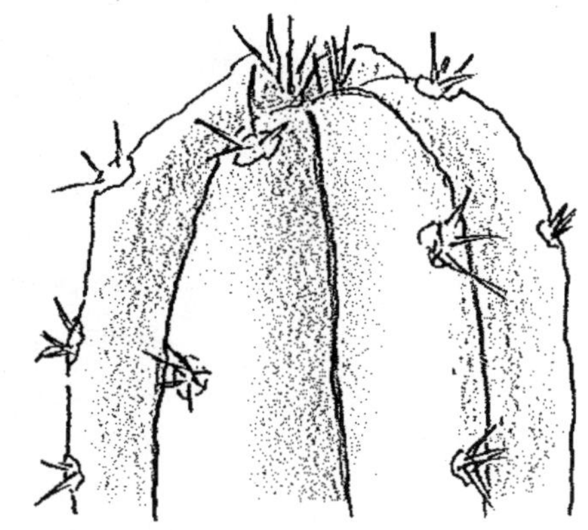

Some plants have adaptions to minimize water loss from their leaves, such as the spines of cacti. The spines are modified leaves.

Some plants have small hairs that cover the surface of a leaf to reduce transpiration. Also shown are the stomata (like eyes) that close during the hot part of a day.

» DID YOU KNOW?

On average, approximately 4,000 gal a year can be saved in homes by replacing inefficient showerheads, installing tap flow control devices and fixing leaks, while a further 5,000 gal a year could be saved through a retrofit of inefficient toilet suites.

Approximately 13,000 gal a year can be saved by replacing a well-maintained lawn (turf) with a waterwise garden.

Irrigation practices

Water shortage and high energy costs motivate gardeners to harvest the greatest possible yield from every precious drop of water.

A waterwise garden can be achieved by either installing an efficient watering system or replacing lawn and other water-hungry exotics with waterwise plants, or both.

Designing your home's irrigation system to be waterwise would not only conserve water but also save you money on your water bills. Most people who have automatic systems rely on sprinklers and sprays. The trend these days is for dripline systems.

In many cases, contemporary spray and sprinkler irrigation systems can be extremely wasteful of water. Some are not very environmentally friendly, they may increase the risk of plant fungal diseases, and they are inflexible and ill-suited to complex garden layouts. Choose sprinklers that produce large droplets and are water efficient.

Furthermore, landscape-sensible design will enable reduced fertilizer use, and reduce runoff and thus soil loss. Coupled with an appropriate use of mulch, organic compost and soil amendments, your garden will still thrive on reduced water use.

A low-irrigation garden should result in:

- reduction in water demand
- minimization of runoff, and an increase in water absorption into the soil
- replacement of scheme (town) water with stormwater, rainwater or graywater
- efficient irrigation of plants
- lower evapotranspiration, by the use of mulches and substrata or subsurface irrigation methods.

Proper watering methods are seldom practiced by most gardeners — they either under- or over-water.

The person who under-waters usually doesn't realize the time needed to adequately water an area and the volume required by growing plants. A light sprinkle on plants is harmful as insufficient water is made available to the plant. Light sprinkling only settles the dust and does little to alleviate drought stress of plants growing in hot, dry soil.

Generally, it is best to water the garden area for a longer time, but less frequently. A good soaking is far more effective. This type of watering allows moisture to penetrate into the soil area where roots can readily absorb it. A deeply watered soil retains moisture for several days. However, the watering regime does depend on the type of soil. For example, you should water more frequently on sandy soils and less frequently on clay soils, but over-watering on deep sandy soils is not advised as the water quickly passes through the root zone downwards and is not able to be absorbed by plants.

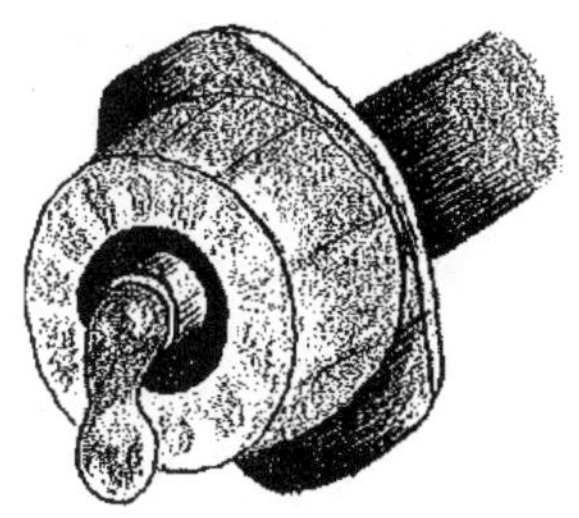

Individual drippers emit a set amount of water each hour.

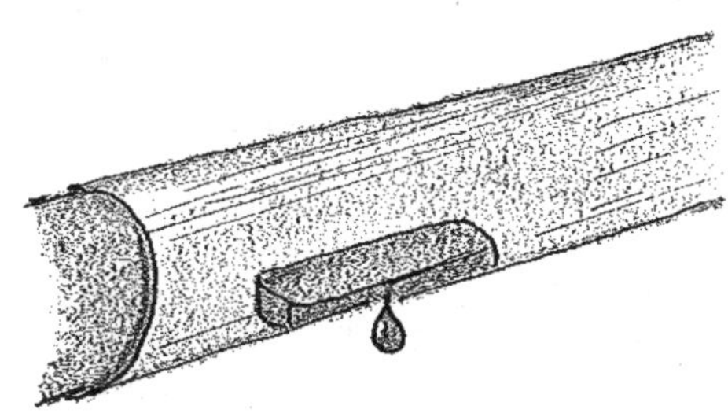

Dripline irrigation contains embedded drippers at regular spacing inside the tubing.

Then there are those people, with the best intentions, who water so often and heavily that they drown plants. Symptoms of too much water include leaves turning yellow or brown at the tips and edges, then brown all over and then dropping from the plant.

Too much water in a soil also causes oxygen deficiency, resulting in damage to the root system. Plant roots need oxygen to live.

Thoroughly moisten the soil at each watering, and then allow plants to extract most of the available water from the soil before watering again.

Tree Height	Gallons of water/ tree/day while in production	Gallons of water/ tree/day for survival
< 6 ft	13	3
6–15 ft	40	5
> 20 ft	100	18

Table 10.1. Water requirements of trees.

Determining the amount of water to apply to trees is probably one of the most difficult questions to answer. Many factors must be considered, such as the soil type, environmental conditions, the nature of the specific crop, the size of the tree and time of the year, e.g. sandy more than clayey soils, summer more than winter and spring, fruiting more than general growth, wind more than still air etc.

You might be surprised to discover the amount of water some trees require, as shown in Table 10.1. Frugal gardeners tend to under-water fruit and nut trees and underestimate how much water they will need each year to have a productive food garden.

The secret is to make sure that plants get the right amount of water, when they require it. This involves monitoring and observing. There are many different devices available for determining the soil-moisture status in the garden. These include tensiometers and gypsum blocks.

» DID YOU KNOW?

Only about 1% of the water a plant obtains and uses is retained in the fruit, and less than 0.5% in the remaining parts of the tree — the leaves, shoots and roots. The other 98.5% of the water supplied to the tree is lost by transpiration.

A typical fruit tree contains about 85% water and 1% inorganic nutrients (fertilizers), while the other 14% is made up of sugars, starches and other organic compounds.

The amount of water used by plants increases:

- as the air temperature rises
- during dry windy weather
- when soil temperatures are high.

Gardening can be complicated, can't it? It doesn't have to be, and we can easily improve our irrigation understanding and skills.

For example, different watering regimes have different efficiencies (a measure of how much water is used compared to how much is wasted), such as:

- surface water use, e.g. furrows 60–75% efficiency (25–40% of water is wasted)
- spray 65–75%
- drip 75–90%.

Why use drip irrigation?

I suppose the more basic question to be addressed is, "Why do we need to change?" Besides the increasing cost of water and power, depletion of our water resources and pollution of our water bodies, there are pressing demands for more efficient systems and a need for safer ways of reusing our graywater and collected rainwater.

Drippers are by far the most efficient way to water plants. Individual drippers typically come in 0.5, 1 and 2 gal/hr, while dripline drippers vary from 0.3–2.5 gal/hr.

Furthermore, technology is such that drip irrigation can be pressure-compensated, self-flushing, anti-siphoning and have a shut-off mechanism. A pressure-compensated dripline may have drippers emitting 0.5 gal/hr while a non-compensating dripper could be 2 gal/hr.

A drip irrigation system is easy to install and is economical, and is not as complicated or as costly as you may think.

Benefits and limitations of dripper irrigation

With any type of irrigation system there are pros and cons. In this case, the pros certainly outweigh the cons. The benefits of drip irrigation include:

- efficient water use (uniform distribution, good recovery)
- low application rate (reduced risk of runoff)
- ideal for odd shapes and narrow strips
- improved disease control
- effluent (e.g. graywater) reuse
- reduces weed growth
- watering over a longer period is possible
- reduces exposure to vandalism
- reduced injury risk
- more energy efficient
- ability to better use fertigation (injecting fertilizer into the irrigation network).

On the other hand, drip irrigation does have some drawbacks, which include:

- requires capillary action of water to work
- more technical maintenance required
- establishment of lawn may require temporary overhead watering.

In some cases, drippers or dripline may need replacing after 3–5 years if they become blocked. You can minimize this by using dripline and drippers that can be flushed, using copper drippers to prevent root intrusion or having a mild herbicide dosing system to stop roots from entering the dripper pores.

The basic rules of waterwise irrigation

- Water in the early evening or night, thus reducing loss by evaporation. About 60% of water is lost if you use fine sprays during the heat of the day — the water either never touches the soil or quickly evaporates as it does.
- Water the roots. How many times have you seen someone watering plant leaves? Leaves do not absorb much water this way — it mainly enters the plant via root uptake.
- Turn irrigation off during the winter months (or whenever you experience your main seasonal rainfall). Generally, irrigation can be turned on from spring through to autumn, and then it should be controlled so that if rain events do occur the irrigation system is not activated.
- Don't water paths and driveways. Overspray is a complete waste of precious water. Make sure your system is set up to deliver the right amount of water, at the right time, to the right place.
- If you must use sprinklers buy ones that produce large droplets to minimize wind drift. Only use micro-irrigation for intense garden beds that need humidifying and are sheltered and protected from wind and sun.
- Don't over-water. Every soil type only holds a certain amount of water, so regulate the volume you add. Even a simple and inexpensive tap timer on manually operated sprinkler systems helps.
- Consider hand-watering, which can be both relaxing and efficient, especially if you have a trigger nozzle that only allows water to leave the hose when it is pressed.
- Over-watering can often occur if the water pressure is high. You can install a pressure-reducing valve to keep the water pressure as low as possible to reduce water wastage.
- Installing a rain sensor will also prevent over-watering. These devices monitor or respond to either rainfall or soil moisture and prevent the controller from switching the irrigation on.

Common types of sensors include tensiometers, gypsum blocks and capacitive sensors. Evapotranspiration sensors are relatively new products that can monitor changing weather conditions, even when there is no rain and adjust the watering regime daily.

Rain sensors vary in price, but are typically about half the cost of a soil moisture sensor.

Heavy droplet sprinklers reduce loss by wind. Notice the overlap of spray.

» DID YOU KNOW?

If you use a high-flow dripper (e.g. 2 gal/hr) much of the water is lost to plants as it tends to move downwards quickly, especially in sandy soils.

When a low-flow dripper is used (e.g. 0.4 gal/hr) it drips slowly enough for the applied water to spread sideways as well, ensuring that water is kept in the top soil, and is therefore more available for plant uptake.

The disadvantages of low-flow drippers are that you typically need a high pressure pump to make them work (as they are pressure-compensating) and secondly the pump is on longer, so operating (running) costs are higher.

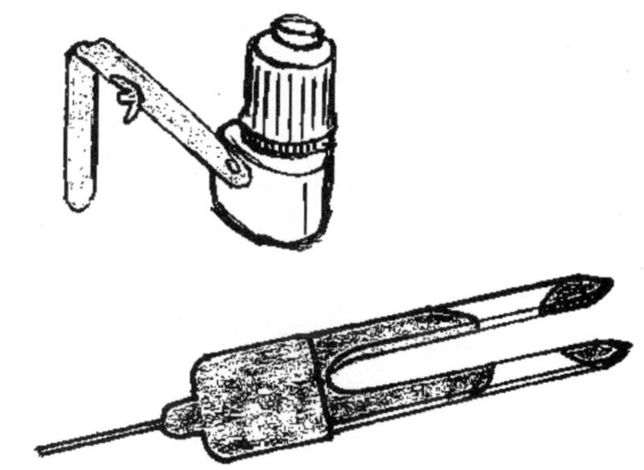

Rain and soil sensors can prevent over-watering.

Graywater reuse

Graywater is the wastewater stream from all sources other than the toilet (toilet water is often called blackwater or sewage). Kitchen graywater (and dishwasher water), however, should not be reused as this can contain oil, fats and food scraps, which do not break down easily and can easily clog irrigation filters and pipes. This means that householders can easily reuse greywater from the bathrooms (shower, handbasins and bath) and the laundry.

Graywater can be reused in a variety of ways, such as watering garden plants and for toilet flushing, which requires the installation of a treatment system. Watering our gardens with graywater can be easily achieved.

Water is too valuable a resource to waste, and any endeavor to reduce freshwater consumption or reduce wastewater disposal and treatment and the energy this consumes should be encouraged.

Why recycle graywater?

Typical values for the volume of daily graywater produced range from 20–25 gal per person, so reusing graywater onto the garden could give a family of three or four an extra 25,000 gal of free water a year.

Reusing laundry and bathroom wastewater on your garden is just another way of moving towards a more sustainable lifestyle. What's more, it's a practical step that is just as easily applied in the suburbs as it is in the country.

Graywater reuse strategies

Graywater can be reused in a number of different ways. These include subsurface drain systems and substrata dripper irrigation for plant irrigation.

There are now many different graywater systems approved for use, but you often need to submit an appropriate application and fees, typically to the local government council. A licensed plumber is required for any changes to the sewerage system.

Graywater cannot be used to irrigate a vegetable garden that contains below-ground food crops such as onions, potatoes and carrots, but can be used on above-ground crops such as tomatoes, broccoli and corn, and on fruit trees, lawn areas and on other plants such as exotic and native shrubs and trees.

Placing graywater in the root zone of plants is the most effective way to ensure maximum uptake of both the water and the range of nutrients that are available in graywater. A word of caution: many Australian native plants are susceptible to high levels of phosphate. These include the family Proteaceae,

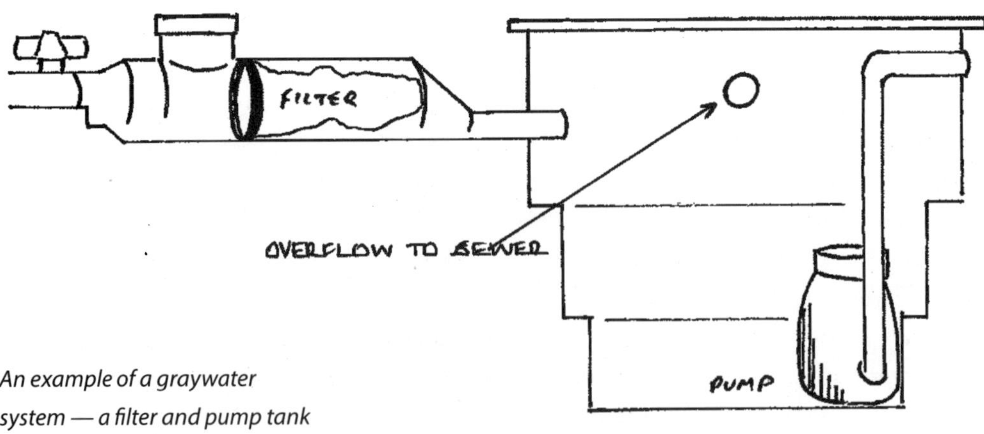

An example of a graywater system — a filter and pump tank

such as grevillea, banksia and hakea. Some introduced (exotic) plants, such as azaleas, camellias and gardenias, do not like the alkaline nature of some graywater sources. It is best not to use graywater on any of these types of plants.

Graywater should first pass through a filter or settling tank before dispersal in an infiltration area. The filter or tank removes coarse material, such as hair, soap flakes, sand and lint, which could block the drippers, draincoil pipe or soil.

Dripper systems are the most common. Depending on the slope, you can gravity-feed the graywater to the drip irrigation or you may have to pump it to garden areas. Pumping water is always more expensive, but it gives you more flexibility, and you can send the recycled graywater to any garden bed on the property.

The Dos

- Select garden friendly detergents. Only biodegradable products and products with low phosphorus, sodium, boron, chlorine and borax should be used. Bleaches and fabric softeners should be used sparingly.
- Apply graywater in several locations rather than one single point, so that pooling of graywater does not occur.
- Apply graywater to areas that are not readily accessible to children and household pets.

The Don'ts

- Don't use graywater from the washing of diapers and soiled clothing.
- Don't use graywater when a household resident has an infectious disease, such as diarrhea, infectious hepatitis or intestinal parasites.
- Don't discharge graywater on edible plants or where fruit fallen to the ground is eaten.
- Don't store graywater. Stored graywater will turn septic, giving rise to offensive odors and provide conditions for microorganisms to multiply.
- Don't over-water, which can result in the development of unsightly areas of gray/green slime. This slime is caused by the presence of soaps, shampoos, detergents and grease in graywater. The accumulation of slime can cause odors, attract insects and cause environmental damage.

Health concerns

Graywater contains a range of pathogens, those organisms that may cause disease. Graywater is not allowed to be used in ponds or for above-surface irrigation systems due to the risk of mosquito breeding and contact with human skin and possible pathogen transfer.

Many pathogens such as bacteria (e.g. fecal coliforms) and protozoans (e.g. Giardia) may be present in some graywater sources.

Graywater also contains bacteria and other microscopic organisms that feed on the nutrients in graywater, causing the wastewater to smell after a day or two.

High levels of nitrate and phosphate may be beneficial to many plants, but can be detrimental to humans if ingested.

Rainwater harvesting

Our rainfall patterns and our climate are changing, and this is all part of a larger global picture, but we all need to plan for changes happening locally.

The average home could easily harvest thousands of gallons each year from their roof, and this water can be used to offset the water budget for your home.

Installing a rainwater tank, no matter what size, is a small step that you could take to become self-reliant.

Generally, the larger the tank the more water you can collect and use. If you wanted to use rainwater to flush toilets and provide water to the laundry, then you would probably need at least a 2,000 gal tank. It is not uncommon to install 7,000–12,000 gal tanks for this purpose, but all of this depends on your annual rainfall patterns.

» DID YOU KNOW?

To calculate how much rainwater falls on your roof, multiply the annual rainfall (in meters) in your region by the roof area of the house. For 1m^2 of roof and 1mm of rainfall, 1l can be harvested, so, for a 200m^2 roof area and annual rainfall of 500mm (0.5m), this equates to 100m^3 = 100,000l each year.

If you use the imperial system, multiply the roof area in square feet by the number of inches of annual rainfall, and then multiply the result by 0.623. For example, a roof area of 2,000 square feet and an annual rainfall of 18 inches equates to 22,500 gallons (about 100,000l).

What are the benefits of rainwater tanks?

Collecting rainwater has many environmental benefits, as well as benefiting you! Some reasons for harvesting rainwater include:

- making freshwater available to flush toilets or to provide a laundry source
- using rainwater for drinking purposes
- supplementing the watering of garden areas
- reducing our use of town water — a very valuable, limited resource
- saving some money — buying less water from a service provider or utility
- providing a water source that has reduced levels of salts and other substances.

Types of tanks

Rainwater tanks are now made from a variety of materials. Generally, the most popular for urban backyards are either made from steel or polyethylene.

Large tanks (12,000 gal plus) are typically steel-liner tanks. These have a steel outer structure with a flexible poly liner inside.

These days, poly tanks are UV stabilized and come in a range of colors. Both steel and poly tanks normally have at least a 15-year warranty.

If there is space below the house or veranda then a bladder tank is an option. These tend to be proportionally more expensive, but can be ideal if there is little room for a regular tank.

Finally, more and more tanks are being buried as new homeowners build larger houses on smaller blocks. Below-ground tanks can be buried under decking provided there is access to service the pump or to clean out the tank if required.

Buried tanks can be one-piece poly or concrete tanks or interlocking plastic cubes wrapped in a poly liner. These latter types can be almost any size and shape.

Some rainwater tank systems are suitable for burial, and the storage "cubes" are wrapped in a waterproof fabric.

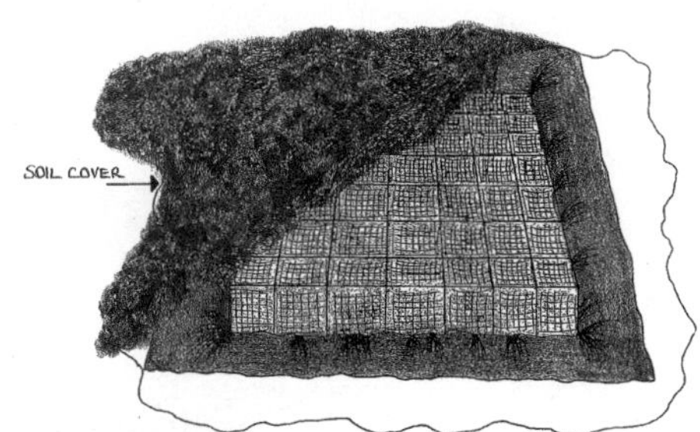

Connections to the house

Most people want to use the rainwater — either for drinking, to flush toilets or to wash clothes. Rainwater is most often pumped to the house, although gravity can be used in some cases to direct rainwater to fixtures in the house.

Either a submersible pump, a pressure-tank pump or a pressure-switch pump is used to supply rainwater when required.

When the tap is turned on, or the toilet flushes, the pump is activated and gently pumps water to fill the tank, or enter the kitchen sink or washing machine.

If you wanted to use the rainwater for watering the garden, then think about how much water your plants actually need to survive.

If you work on 0.5 in of water every fourth day applied to the garden then you can calculate how much water you require. For example, if the garden bed is 400 ft^2 (say 20 ft × 20 ft) then you require 100 gal. So, a 200 gal rainwater tank would only last for one week (two waterings).

It is always best to install the largest tank possible, keeping in mind the size restrictions you may have at your home and the cost. You don't really need to water the garden in winter, so you should work on summer watering only.

» **DID YOU KNOW?**

Using 0.5 in as the preferred plant watering regime, this equates to 2.5 gal of water for every 10 square feet of garden.

What happens when I run out of rainwater?

If you only install a small tank (e.g. less than 5,000 gal) then it is likely you will run out of rainwater during the summer period. This, of course, depends on the uses of the rainwater and whether your area experiences summer rain to top up the tank. Providing a full laundry, kitchen and bathroom service rapidly depletes the volume you can collect during rainy times.

Rainwater tank systems can be set up to integrate the town water source with the rainwater source. An automatic water-sensing device enables town water to enter the system when the rainwater is depleted. There are several ways to achieve this, from manually changing valves to fully automated switching devices. It is important to always have water to flush toilets and to provide a source to the laundry.

What you should also consider

Most rainwater tanks come supplied with a basket (leaf) filter, tap and overflow pipe. In addition to these standard fittings, a number of optional extras are available for your rainwater tank system. These include:

- First-flush device. This enables the first rains to be directed away from the tank. This water may contain dust and decayed matter, and it is best not to collect this and pollute the tank water.
- Vermin proofing. This is often necessary for steel and steel-liner tanks to prevent insects, frogs and small rodents from finding their way into the tank.
- Garden overflow. Either a subsurface piped trench or a simple gravity-fed dripper system is installed to direct overflow more effectively to garden areas or beds.
- Leaf eater. This is a screen that filters rainwater and allows leaves to be shed from the system.

A typical rainwater tank setup includes a first flush, leaf eater, water-switching device and overflow pipework.

What are the costs involved?

Rainwater tanks are relatively cheap. However, small tanks are proportionally dearer, so the larger the tank the more cost-effective it is. If you intend to pump rainwater to flush toilets and so on, then a pump and irrigation filter would be required.

Installation would be extra, and this depends on the distance to the house fixtures and the degree of difficulty in supplying water to the house.

Adding options such as a leaf eater, first-flush device, overflow to gardens and a filter bag is highly recommended.

» DID YOU KNOW?

If each household installed a rainwater tank, reused its graywater, modified the garden, installed water-efficient fittings and appliances, and minimized water wastage, then we could reduce domestic water consumption by 50%.

Integration of water systems

Installing a graywater system or rainwater tank is commendable but when you integrate these types of strategies as part of a water management plan, then real differences in water consumption and use can be observed.

Imagine harvesting rainwater, using it to wash clothes, and then collecting the graywater to use on the garden. In many households 50–60% of municipal supplied water is used inside the house and the balance outside in garden areas.

Laundry use can be anywhere from 14–18% and toilet flushing 12–15% of the total daily use, so that totals just over a quarter of our internal household use. Bathroom use is even greater at about 18–20%.

Now if we can use rainwater to completely supply laundry and toilet flushing, and then capture all of the laundry and bathroom water by a graywater device to water the garden, then we have reduced our town water use significantly.

If we can reduce town water supply by 25% and then capture 33% of our household use to offset what we use in the garden, then we greatly reduce our overall consumption. We can then further reduce our outside water use by installing dripline or some other water-efficient watering equipment.

This holistic approach of an integrated water system is the concept behind town water neutral gardening, as advocated by Josh Byrne. Through design and the use of water-efficient devices and techniques, appropriate fit-for-purpose water sources can be better utilized. For example, it seems absurd that we use disinfected class A water (drinking water) to flush a toilet, and if shallow well water is available then it can supplement graywater use in the gardens when the family is on holidays or at other times.

We need to promote improved water management through behavioral change, and this then leads to reduced water consumption. Rural families that solely use rainwater for their water supply generally use 15% less water than their urban counterparts.

People are misguided if they think that technology will overcome future water and resource shortages, so getting people to change what they do and how they think is the battle we face.

On-site domestic wastewater systems

Wastewater is really wasted water. It is a resource we cannot afford to literally throw away. There are three main ways in which our daily household wastewater is treated. Many urbanized and modern cities have a reticulated sewer system, whereby household wastewater is piped to a central treatment plant and then pumped to discharge — either to the ocean or on land.

This is an expensive way to treat wastes, simply because of the huge investment in infrastructure, operation and maintenance that is undertaken for these large-scale projects.

» DID YOU KNOW?

The aim of sewage treatment is to remove or reduce the following pollutants:

- organic matter (monitored as BOD — Biochemical Oxygen Demand)
- suspended solids
- nutrients such as nitrogen and phosphorus, too much of which negatively impact on our waterways
- pathogens (organisms that cause disease).

If household wastewater is not piped off the property then some type of on-site treatment plant is required. The first, and cheapest, option is the septic tank. This can be two tanks (or one tank with a baffle to separate two chambers or compartments) that are essentially an anaerobic digestion system (hence the term "septic").

The first tank or chamber is the primary sedimentation tank. It holds all of the solids from the household, and much of this is broken down by bacteria and other microorganisms. Treated effluent is drawn from the middle of the first tank and enters the second tank where further digestion of the waste occurs.

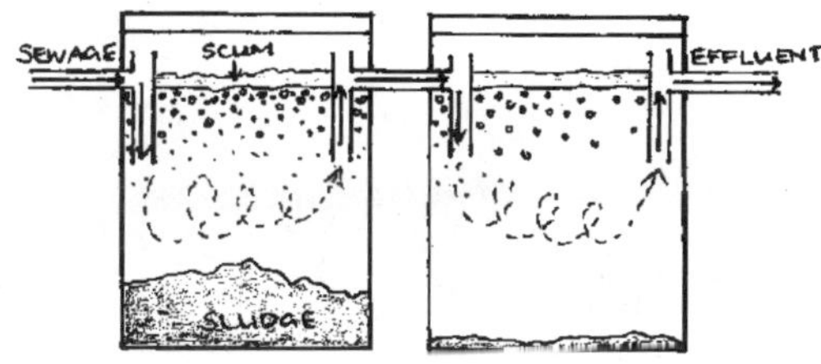

Double septic tank system

Septic tanks are designed to hold all of the daily waste for a few days to enable break down and separation. The retention time that waste is held is crucial to the success of waste digestion, and if too much wastewater enters the system then the bacteria don't have enough time to break it all down.

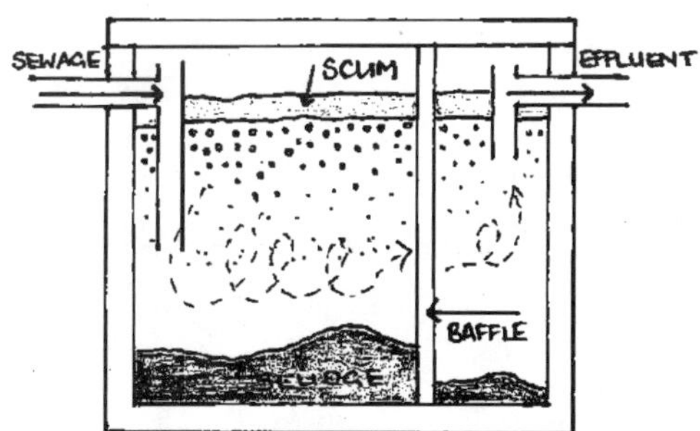

A single septic tank may have a baffle to separate chambers.

Partially treated effluent is then passed into some type of soil absorption system. This may be leach drains, stone-filled beds or trenches, or soak wells. The size of the leach field is proportional to the daily volume of wastewater and inversely proportional to the porosity of the soil. For example, if the soil is clayey in nature and water only percolates slowly into it, then the drains need to be larger or longer.

All wastewater treatment systems require maintenance. Unfortunately, most people ignore the septic tank until it floods or starts to smell or the toilets back up and a plumber is called. Once the tanks and drains are buried most people have an "out of sight, out of mind" attitude.

In non-sewered areas, septic tank systems are still the norm. However, sometimes the soil is not suitable for continuous discharge and there may be sensitive waterways nearby or a high water table. As more councils require stringent effluent discharge quality, the installation of secondary treatment systems is steadily increasing. These whole-of-house systems treat all wastewater to a much higher standard and the discharge water can then be used to irrigate gardens.

An alternating leach field

An Aerobic Wastewater Treatment System (AWTS) is also called an Aerobic Treatment Unit (ATU) and they provide both primary and secondary treatment of domestic wastewater. Typically, they consist of several chambers (or even separate tanks) where some combination of biological, physical and chemical processes is employed to remove the pollutants.

The first chamber is the primary sedimentation chamber and it operates in a similar manner to a septic tank. Anaerobic bacteria digest much of the wastes and produce various gases such as carbon dioxide, methane and nitrogen, which are vented.

The second stage is the aeration stage. A blower pumps air, either continuously or on a cycle, through a diffuser, which forces air bubbles into and throughout the wastewater effluent. Different types of bacteria exist when air is plentiful and the chemical processes that occur are different also.

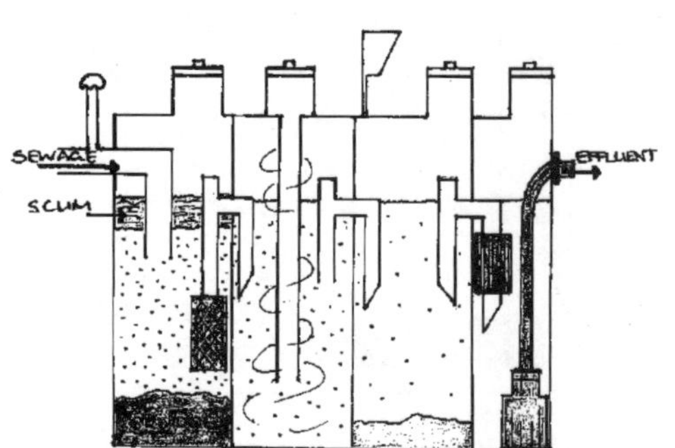

A cross-section of an AWTS

Once all of the effluent has been stirred up it needs time to allow the floc (minute undigested or insoluble particles) to settle to the bottom of the clarification chamber as sludge.

Every septic tank and AWTS will accumulate scum (floating on top) and sludge (muck that sinks to the bottom). This is because the bacteria that digest human wastes are not that good at breaking down oil and fats (scum) and some solids (and this includes toothpicks, cigarette butts, plastic, condoms, cotton buds and lumps of toothpaste, which unfortunately end up in the system).

Much of the sludge is also dead bacteria. Eventually every tank needs to be pumped out and desludged, most often between 3–10 years, depending on the system and what enters the system.

Some of the sludge in the clarification chamber is returned, often by an air-driven lift pump, back to the primary chamber. Monitoring the volume of sludge in the aerated mix is also a good indication of when the tanks need pumping out by a liquid waste contractor.

The clear, settled liquid is then passed through some type of disinfection process, using chlorine tablets, UV light or ozone, and then into the pump chamber. As the treated water rises in this chamber, the float switch activates and water is pumped to the irrigation area.

All of these very complex processes are designed to treat sewage to a level suitable for surface irrigation. This includes dripline or special sprinklers, which only permit a short width and height spray so that the treated effluent is not showered on buildings, paths, animals and people.

Every state, province or country has rules and regulations about how much irrigation is required, plume (spray) height and the setback distances from buildings and boundaries.

Aerated treatment systems are biological systems and require servicing and maintenance by experienced, registered service technicians. They are more expensive than septic tank systems to install and maintain.

Furthermore, pumps and blowers use electricity and occasionally need replacement. All of this adds ongoing costs to the overall operation of the system, but about 50,000 gal each year can be used to irrigate lawns and gardens.

» DID YOU KNOW?

Most municipal wastewater treatment plants are designed for tertiary treatment. This is a further stage of wastewater processing before any effluent is released to the receiving environment. Here, more nutrients are removed (so less nitrogen and phosphorus), greater filtration and settling may occur to produce higher-quality effluent and greater levels of disinfection ensure complete pathogen kill.

Water in the landscape

Besides capturing, using, treating and recycling water, we need to keep water in the soil. This is particularly important on rural properties (see next chapter), but is easily applied in small urban properties too.

The first step is to capture runoff, and then we choose to hold water there or move it through the landscape and direct it to dams and storage. Basically, we need to slow it, spread it or sink it. How soil can be improved to hold more water was discussed in chapter 4 and moving water through drains and into dams is discussed in chapter 11.

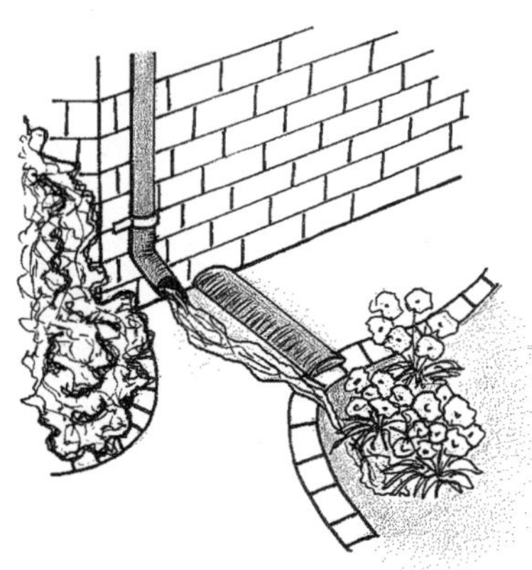

You can divert stormwater off a path into the garden.

On an urban block you could easily divert stormwater from a downpipe into a garden bed, an orchard area or a small basin.

You can create mini compensation basins or sumps in garden areas. These can receive the runoff from hard surfaces (paths, driveway and roof) and soak away over time. These shallow sumps become landscape features in the garden and are an alternative to the more traditional underground storage tanks.

» DID YOU KNOW?

Water has remarkable properties. When it freezes it expands, becomes less dense and this is why it floats.

Freezing water inside plant tissues can have devastating effects. Some plants that live in very cold climates have special features to prevent damage to their cells.

Certain plants seem to be able to change the structure of their cell membranes and as water freezes it can expand through the membrane into other spaces.

Other plants convert starch into sugars and this changes the composition of the liquid inside the cells. Essentially the sugars act like antifreeze and the plant withstands very cold temperatures without the cell liquid actually freezing. While the fluid inside the cell doesn't freeze, any water between cells can freeze and expand, but this doesn't cause much damage and the plant can recover.

Finally, a few plants can allow the cellular liquid to become viscous (thick), almost becoming solid, and this also helps plants to withstand very low temperatures.

STRATEGIES FOR RURAL PROPERTIES

11

BY NOW YOU HAVE REALIZED that we need a lot of tools and skills to help us with the regenerative work on the soils we have. Farming is a complicated process, and many people have misconceptions about it and don't appreciate how difficult it can be at times.

I would like to think that most farmers are conservationists — they live on the land and derive their livelihood from it. They need to manage their properties well in order to make farming a viable business.

So it's in their best interests to care for the land. Even so, some farmers, who have inherited European ways to farm our land (e.g. plowing to turn over the soil), also have little understanding of how to farm in a land-healing way.

Farmland throughout the world is increasingly becoming subject to dryland salinity, soil acidification and loss of soil carbon. These all affect productivity and ultimately the sustainability of the farming enterprises.

Farming is so diverse. It ranges from low numbers of cattle grazing on vast areas of native grasses in rangelands, to cereal and stock rotations in the "wheatbelt," to intense feedlots and battery hens and animals.

I think most people understand that many historical farming practices have had a negative impact on the environment, the land and biodiversity. Some of these practices were misguided and farmers were fed misinformation from institutions, other businesses and government bodies. It is not all their fault!

Over the last century, conventional farming practices tended to be large monocultures of specific crops and often in association with particular animals, such as wheat and sheep. Another issue is that, in most industrialized countries since the 1950s, many farmlands have increased in size while the number of farm owners has decreased, and more and more work is being done by ever-increasing use of larger machinery. In some regions of the world, crop farms are expanding and rangelands are decreasing. Ideally we need the opposite to occur. In today's economic climate it makes sense to engage in lots of different enterprises to create business stability.

Up to the advent of fertilizers, pesticides, herbicides and mechanized machinery, agriculture was undertaken organically with the use of animal manures as fertilizer, simple mechanical (physical) pest control and the occasional encouragement of biological control with beneficial organisms.

What we call modern agriculture really developed post World War II, and the current practices of chemical use, tillage and large-scale monocultures, with little regard to soil, erosion, waterlogging and salinity, have continued to grow.

A small number of farmers are doing things differently, and there has been a slow conversion of farmers to more holistic methods and practices based on organic farming, or to methods that build soil rather than denuding the landscape.

While conventional cropping practices require that all weeds and vegetation be killed (usually by herbicides) before sowing and during crop growing, there are many farmers who now practice no-till, minimum till and a whole host of different techniques that build soil. Building soil, farming based on the principles of ecology and better farming practices became the basis for "sustainable agriculture."

» DID YOU KNOW?

The term "sustainable agriculture" was coined by Australian agricultural scientist Gordon (Bill) McClymont in the 1950s. He was a pioneer for his work in integrating soil fertility, animal husbandry and agronomy (crop and pasture production) to improve both livestock and agricultural production.

Here is a brief appraisal of some of the better-known, "alternative" farming and land management practices. In essence, farming is all about the soil, water management and the types of organisms we cultivate and nurture, and should be based on biological and ecological sciences.

Organic farming

The most important elements in the soil are carbon, oxygen and hydrogen — the same three that are the backbone for all life on Earth. If we take care of these in the soil, we are in a good position to create soil life. We may have to adjust levels of other nutrients such as nitrogen, phosphorus, potassium, sulfur and calcium, but these can be managed by our planting regime and our use of amendments and composts, and with the help of the myriad of micro- and macro-organisms in healthy soil.

Essentially, organic growing and farming uses no synthetic chemical fertilizers and pesticides, and instead emphasises building up the soil with compost additions and animal and green manures, controlling pests naturally, rotating crops, and diversifying crops and livestock.

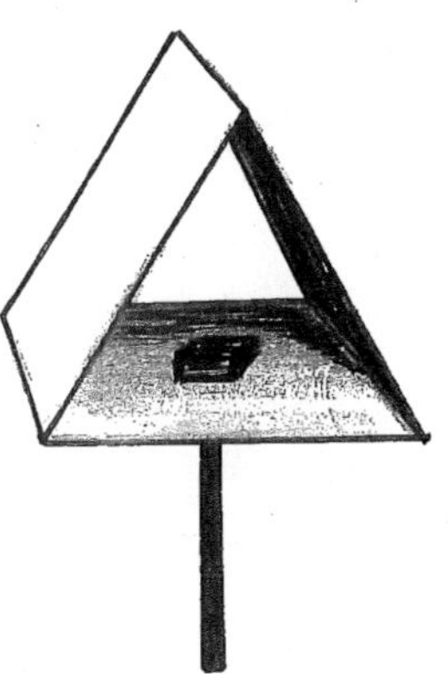

Pheromone sticky traps. Just one strategy for pest control.

Organic gardens, pastures and orchards are both complex and holistic. It is not as simple as using compost instead of soluble nitrogen, phosphorus and potassium (NPK) fertilizer and garlic sprays instead of Malathion. Like permaculture, it is about taking responsibility for your own health and well-being, your family's health and the health of the planet.

Organic is now mainstream and is recognized for its role in issues of environmental protection and repair, and the number of people (including farmers) converting to organic production is steadily increasing.

The initial interest in permaculture and organic growing arose from the undisputed evidence linking the use of common garden chemicals with human health risks and dangers to beneficial creatures in our environment.

Instead of using chemicals, organic growers attack weeds with heat and fire to incinerate them, or spray them with vinegar to destroy them naturally. They fight garden pests with sticky traps and insecticidal soaps, hand-pick pests off plants, and use sprays containing natural products like neem.

» DID YOU KNOW?

Pesticide poisoning affects 48 people per minute (25 million every year worldwide). Homeowners use more pesticides per acre than farmers.

A word of caution: some organic gardeners contend that natural pesticides such as rotenone (derris dust) and pyrethrum are mild, safe and suitable for household use. However, much research has revealed that rotenone is toxic to mammals and fish and produces symptoms in humans similar to those of Parkinson's disease. Pyrethrums have been found to produce tumors in animals and are toxic to fish.

So, let's try to reduce our use of all chemicals of this nature, and recognize that some are more deadly than others.

The basic principles to grow your own organic produce include:

1. Building up the soil to be high in organic matter, nutrients and water-holding capacity.
2. Controlling weeds and plants that harbor pests and disease.
3. Rotating crops and using companion planting.
4. Encouraging beneficial insects and other animals for pest control.
5. The placement of plants is crucial (it's all about design!). Tall sweet corn might shade struggling broccoli, and you need to think about sun angles, shade and microclimate when positioning plants in the landscape.

Growing food is relatively easy. It is simply just practice and practice, and, as you know, practice makes progress. You make mistakes, plants die, they get eaten before you get a chance to, you under-water, over-fertilize, encourage pests rather than controlling them, and don't have the time to spend in the garden as much as you should. However, the enjoyment of growing, picking and eating your own food is exhilarating. Benefits of organic growing include:

- getting the nutrition you need
- enjoying tastier food
- food is fresher
- saving money
- beautifying your community
- fewer food miles
- protecting future generations
- protecting water quality
- preventing soil erosion
- saving energy (less transportation and pollution)
- promoting biodiversity
- using less chemicals (most herbicides, fungicides and insecticides are thought to be carcinogenic).

Organically grown fruits and vegetables may cost a little more at the store, but the great flavor and health benefits are well worth it.

Many nutrients are often found in higher levels in organic foods, but the concentrations of each do depend on what the grower is supplying the plants with.

Organic production is slower and generally has lower yields than conventional methods. Much of mainstream agriculture is subsidized in some way by governments or industry, or both. This is why organic food could be up to 60% more expensive.

Organic growers face an uphill battle. It is labor intensive growing and getting quality laborers is just as demanding.

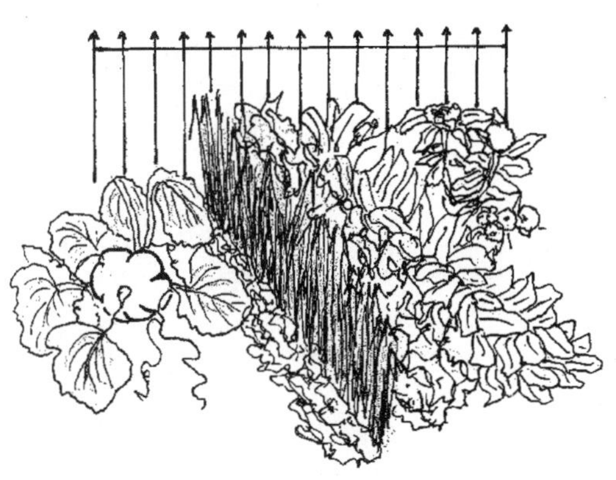

Organic growers mix plants up to minimize pest attack.

You also need a real understanding of the optimum growing conditions for each crop. Backyard growers simply put in tomatoes along with zucchini, watermelons and other summer crops, but tomatoes struggle in temperatures above 79°F whereas zucchini thrive in over 86°F. (As a comparison, summer crops such as tomatoes prefer a growing temperature of 75–79°F, bananas 88–90°F and zucchini 91–95°F, whereas winter crops are much lower — broccoli 70–75°F and lettuce 61–64°F.)

These subtle differences could contribute to the viability of the commercial organic grower where optimum production is the key to success.

Important differences are found between the biodiversity on organic and conventional farms, with generally substantially greater levels of both abundance and diversity of species on the organic farms, as the lack of herbicides and pesticides encourages wildlife.

There is also lots of debate about open-pollinated seeds and hybrid seeds. Over the years it had been drummed into me that we should only be saving heritage seed and open-pollinated varieties as hybrid seeds were bad, didn't produce true types and the seed was poor or even sterile.

Having seen the growth of these different seed types it is clear to me that it is foolish to be so narrow-minded about seeds. I've seen hybrid plants much healthier, not getting mycelium wilt and other pest attacks, and producing much more food than the adjacent, open-pollinated varieties.

And it is difficult to ensure that the seeds you save are open-pollinated and not contaminated. Unless you are isolated from neighbors, it is unlikely you will always get pure seed. Beetroot easily crosses with silverbeet and other chards, and you get colored varieties such as "rainbow chard" (you can buy this from nurseries).

Biodynamic farming

Developed by Rudolph Steiner in the early 20th century in Austria, biodynamics (also called anthroposophy) embraced agriculture and food production by considering spiritual, ethical and ecological philosophies. Essentially, biodynamics views annual crops and soil as a whole system, and it utilizes various additives to activate and enrich soil, as well as the use of an astrological sowing and planting calendar.

Biodynamics is organic growing with a few twists. It relies on using particular sprays made from compost, minerals or herbs, or some combination of these, and planting at particular times based on the phases of the Moon. Biodynamic farmers add nine specific preparations (numbered 500–508) to their soils, crops, and composts to enhance soil and crop quality and to stimulate the composting process.

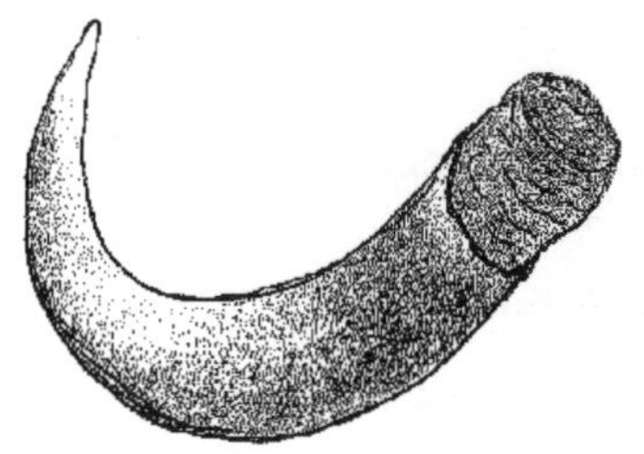

One biodynamic preparation — manure in a cow horn

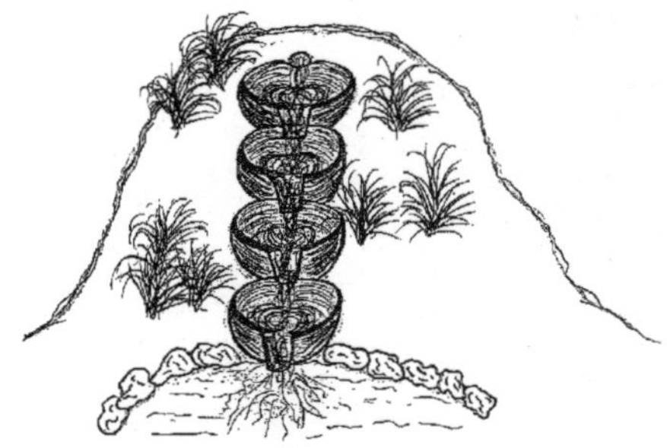

Flowforms are used by biodynamic practitioners to treat and enhance the quality of the water.

Various herbs, such as yarrow, chamomile, stinging nettle, dandelion and valerian, are fermented or juiced as part of the preparation process. Cow manure is also used as one of the preparations, which is used as a soil stimulant before planting.

The second significant difference is the belief in the influence of the Moon and planets to enhance germination, growth and production.

The effectiveness of biodynamics as an agricultural method is still unresolved. Many studies have shown improved soil life, greater numbers of earthworms and plant nutrient uptake, while, generally, producing lower crop yields than conventional farming. More refereed scientific research needs to be undertaken to fully understand how biodynamic preparations influence the soil and plants.

However, as biodynamic produce generally commands higher premiums, the profitability for the grower is often greater than conventional farms. Having said that, not everything has to be measured in economic returns. There are also environment and soil health considerations, and clearly organic and biodynamic farming techniques do improve these aspects, while conventional farming practices have led to degraded and depleted landscapes.

Holistic Management

Holistic Management is a decision-making framework whereby decisions are made based on environmental, economic and social considerations.

It uses a series of steps to enable you to best utilize the management tools available, and along with your observations about ecosystem health, to make decisions about the management of resources and the quality of life you wish for your family and enterprise.

The idea behind Holistic Management was based on Allan Savory's observations of herd animals in the 1960s, which moved throughout savanna grassland in southern Africa. He noticed that dense herds moved each day, pruning the grasslands, and then depositing tons of manure in that area. Birds moved in to pick through the insect grubs in the manure, thus breaking up the pads, tilling the soil and minimizing pests.

All of this was essential for the health of the grasslands and he applied what wild herbivores did in nature to domestic livestock, recognizing that changing one aspect of a system affected other parts of the system, and that each environment was unique. Subtle changes to the time herds were moved or areas allowed to recover were adopted, and it became clear that properly managed livestock was ultimately crucial to the health of the land.

Savory recognized that time was more important than numbers, and it was not about the time you graze but rather the time given to plants and the soil to regenerate and recover after grazing. The secret to the possible effects of herbivores on pasture is management — when to move the animals, how long the grasses need to recover and so on. The movement of animals is, therefore, time-controlled, and adjustments can be made depending on the rate of plant growth, which varies from species to species. Holistic Planned Grazing is a management tool that fits within the Holistic Management framework.

Grazing farm animals are herbivores. But it is not a simple matter to just let them roam about the paddock, as this leads to poor soil and plant

Electric fences keep stock in confined areas.

Stock are kept close together for a short period of time.

management. Herbivores in nature occur as mobs and are continually moving in search of food. If we take this approach at the farm level, we need to keep them tight together for short periods and then move them along. Electric fencing enables farmers to control and move stock in much the same patterns in nature.

Herbivores prune to enable a rapid growth response by plants, and this is what we want. Grass is generally the highest energy converter of all plants, followed by shrubs and then trees. Herbivores have evolved to ferment the grass in their gut and release a number of substances that maintain them. So we need to give them enough time to eat and then rest while the fermentation process occurs. Herbivores eat a lot of plants. One cow can typically eat 26 lb or more each day (dry weight), a sheep about 4.5 lb while a horse eats 22 lb. These figures are expressed as dry weight as the water content varies in foods. It also depends on the size of the animal and whether it is a dairy or beef cow, for example, and so on.

» DID YOU KNOW?

Pasture builds soil much faster than tree crops, but only perennial grasses, and when there is sufficient water and frequent manure to add carbon. While grasses have faster photosynthetic rates than trees, trees remove and store much more carbon than grasses. Grass generally does proportionally better when low carbon dioxide levels prevail. If carbon dioxide levels continue to rise, trees will be favored, and grasslands will undergo succession towards forests.

In nature, birds follow herbivores, and they mix and turn over the soil and thus stimulate vegetation. So the ideal situation for a farmer is to have a small group of herbivores followed a day or so later by a larger group of birds. On a farm, chickens are best for this.

Cell grazing

Cell grazing is a variation of strip grazing and rotational grazing. Although some people don't distinguish between these terms, there are subtle differences.

Rotational grazing (as developed by Andre Voisin) is moving stock from paddock to paddock, but the paddocks may still be continuously grazed — the animals are not confined to a very small area and they can wander over a larger paddock.

Strip grazing is a strategy employed mainly in winter where animals are confined to a "strip" of paddock for a short period of time. The pasture is rationed to the animals, so a lush area of paddock can hold a higher stocking density than normal.

If animals were allowed to roam then much of this luxurious growth would be trampled or spoiled by dung and thus be wasted.

Rotational and strip grazing focus on plant and animal production, whereas cell grazing is a more holistic approach that focuses on ecosystem sustainability while still optimizing farm profit.

Cell grazing can be thought of as a more intensive form of rotational grazing. Paddocks are divided into smaller areas (cells) and animals moved from one section to another. It has long been recognized, since Allan Savory's observations of herd animals in Africa, that pastures need moisture, sunshine and rest to recover from grazing. Since farmers have no influence on rain or sun, at least they can control the "rest" time the paddock needs to recover and regrow.

Some grass species need a couple of weeks to recover before they start to grow, so leaving paddocks untouched for three to four months is the minimum rest period.

Animals should not be allowed to graze the plant material to the ground. There should still be sufficient leaf height above the ground so that plants can use sunlight to make the many compounds needed to produce new tissue.

It is far more energy efficient for this process to occur rather than wasting root reserves to enable plants to recover.

So the overall thrust of cell grazing is high stocking rate in small areas for a short period of time (typically one to three days depending on the size of the cell).

Advanced cell grazing

Developed by Geoff Lawton, this adds other dimensions to moving and utilizing stock. Geoff builds a fenced laneway around the property to move

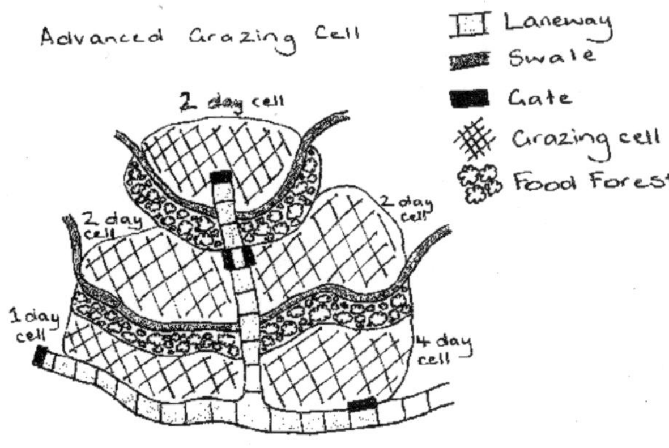

Advanced cell grazing

stock to various "cells" or paddocks, as well as incorporating swales and drains to collect manure washed away in runoff from rain events and directs this to plants in adjacent areas. This fertilizer then increases the growth and productivity of nearby pastures, forests or orchards.

Whereas cell grazing has traditionally been animals on grass pasture, advanced cell grazing can incorporate sections of forest, specific fodder areas, grass and any other combination of these. This gives animals the diversity (and nutritional variety) in their forage, and ultimately, you would think, better health seen in animals.

Polyface

Polyface Farm is the property owned and developed by Joel Salatin and his family in rural Virginia. Salatin extended Allan Savory's idea by using different animals in sequence. Cows initially fed on pasture for a day or so and then they were moved. Chickens were brought in a few days later to feed on grubs in the manure.

Polyface Farm is very animal-centric, with pastures being the main plant focus. Besides cows and chickens, pigs, rabbits and turkeys are used as part of the farm management process. Timber is also harvested from their forested areas. Some other philosophies, such as permaculture, promote a wider range of plants, which normally take up much more of the farm design.

Salatin describes the forage as a "fresh salad bar" as animals are allowed access to a concentrated area of fresh grass so that grazing occurs in a small area over a short period of time. Thus Polyface is more akin to stacking in time as the animals' rotations are scheduled. The overall principle is simple — using mobbing, mowing and moving, with long intervals in between times of very controlled disturbance.

Salatin also recognized that grain should not be fed to stock (herbivores don't naturally eat seed), and if this occurs it is an indication that the pasture is inadequate, deficient or inappropriate in some way. If we grew the right types of plants and fodder trees, there wouldn't be an "autumn feed gap" as feed would be available throughout the year.

As we discussed in chapter 9, not all stock animals eat every type of fodder plant. Some animals are very particular about what they eat, others less so, and some animals have to become accustomed to eating something different.

Chickens follow larger mammals in the paddocks.

Polyface lets animals do what they are good at — their natural abilities and habits can replace machinery. Pigs are arguably the ultimate disturbance tool: pigs are great tractors, plowing land and clearing weed infestations. Chickens can remove both weed seeds and insect pest larvae from the soil, and earthworms aerate the soil as they burrow and produce amazing fertilizer as they digest organic matter.

Furthermore, large animals produce large volumes of manure. A small herd of 100 cows can produce maybe 2 t/ac (dry matter) over the course of a year — more than enough to drive the nutrient cycles that plants require for optimum production.

By stacking complementary enterprises on the land base, you can increase farm profitability. But it all has to do with timing. The initial stock have to be moved well before there is bare ground.

Cows mow down grass to about 2 in but prefer to eat when the grass is about 6 in high, sheep eat down to about 0.5 in, while horses can eventually pull the whole grass root system out. Grass recovery can be as little as three months up to about six months, depending on the variety, soil type, climate and season, and both overgrazing and undergrazing result in poor grass production.

Grass recovery is aided when secondary animals such as chickens scratch and perform other work to break up and spread manure, eat weed seeds, parasites and insect pests, and aerate the soil.

With thoughtful management of pastures, supplementing perennial grasses with a range of fodder and medicinal trees, ensuring the right stocking density and providing enough water for the stock, there may not be a need to buy in additional feed such as grain or hay.

Polyface farming, along with cell grazing and strip grazing, is very labor intensive. Continuous grazing (or set stocking in some countries) uses minimum labor, but moving fences and stock every day, along with bringing in chickens or other animals, makes a very long day for farmers who practice intensive grazing techniques.

Polyface has also pushed boundaries by marketing their organic produce to the local community. Their mantra is "beyond organic" and they sell as much as they can, and at times cannot meet demand.

» WORDS OF CAUTION

Every new idea is always knocked by the establishment. Little scientific study has been made on the ideas behind Holistic Management and Polyface, for example, and hence there are detractors and critics. For example, much of the work and promotion of Holistic Management and Polyface farming, as leading examples, focuses on cows. Occasionally there are pigs and chickens in the mix, but little work it seems on sheep or other animals using these methods on farms. If we can develop systems that do not rely on any additional supplementary feed, other than providing those plants grown on the farm, then much can be said in favor of beef or milking cattle. It has long been the belief of conservationists, vegetarians and many scientists that we shouldn't be growing food meant for humans but feeding animals. Nor should we encourage the clearing of forests and rainforests so that farmers can provide beef to fast food chains. Certainly these pasture-based systems are not managing their farms to the same extent as Savory, Salatin and others.

What needs to occur are more studies on stocking rates based on residual forage. Enough plant material has to be left to ensure soils are protected from erosion, the effects of climate variability on plant recovery (drought) and enough capacity in the plant community to withstand adverse conditions and provide habitat and food for wildlife. Some studies in North America and Australia have produced nonconclusive results and the outcome of the comparison between different agricultural methodologies is unclear. At times, stocking density seems to be of more importance than grazing method, but more work is required.

It is not a simple matter to rotate animals every three months or longer, as plant recovery and regrowth is often seasonal, temperature and rainfall dependent and at times erratic. In some agricultural areas you may have to rest the paddock for a year or more. What is clear from Savory and others doing similar work on the ground is that every farm is unique and has to have its own management plan. Unfortunately one size doesn't fit all.

Natural Sequence Farming

Logs and rocks slow water and allow it to move outwards across a landscape.

Natural Sequence Farming was developed by Peter Andrews, who noticed that if water was slowed down, it spread across the landscape, infiltrating the soil.

As a result, water remains in the soil longer in drought conditions. The additional water infiltrates and settles on top of saline waters below it, thus minimizing their detrimental effects on the landscape.

Water gently moves through a series of ponds where it can slow down and recharge the soil.

While not necessarily organic in its practices, Natural Sequence Farming recognizes weeds as pioneer plants, which were allowed to proliferate and hold and build soil.

After these are slashed to add fertility to the soil, palatable grasses become established.

It is not always possible to change the course of a river or to influence water flow, certainly not over just one farm, so to be effective Natural Sequence Farming has to be adopted by a catchment where all farmers bring about change.

A similar set of techniques, called Regenerative Water Harvesting, was developed by Craig Sponholtz and his colleagues in northern New Mexico.

They developed ways to reduce erosion and restore natural water movement in the landscape, by harvesting runoff in simple structures that allow water to spread throughout the soil, encourage plant diversity and thus build soil.

Craig's work focuses on using both modern river restoration techniques and ancient farming practices to restore (agricultural) land productivity by creating beneficial links between agricultural systems and natural ecosystems.

He uses rock mulch to slow water, prevent erosion and direct water into other areas, and then seeds a large variety of plants to continue with the soil restoration process.

Regenerative Water Harvesting is more of an ecological approach whereas Natural Sequence Farming is more of an engineering approach.

Pasture cropping

Based on the work of Colin Seis and Darryl Cluff, pasture cropping advocates sowing cereal crops directly with native perennial pastures, combining grazing with cropping.

Paddocks are never plowed, so that the native (or exotic) grasses can become established and stay that way. The aim is to have one hundred percent ground cover, and this increases the soil carbon and biomass content and the soil life. Some devotees still use herbicides to control weeds and conventional fertilizers to change the soil chemistry.

The decision to introduce pasture cropping should be based on soil, climate and pasture characteristics. Summer grasses respond to summer rain and many places have indifferent rainfall patterns. Not many places have good summer rainfall.

Growing a winter cereal in a winter grass may not have advantages; growing any plants without the addition of fertilizers of some type to replace plant uptake and loss will only cause rapid decline of the system, and there is less advantage for properties that have higher clay levels in soils that hold onto water for a longer time.

When cereals are grown in pasture there is a yield penalty. Production is lower because some nutrients, such as nitrogen, are utilized by the grasses and are not available to cereals.

On the other hand, if you were not dependent on the income from cereal grain production there are increases in forage for stock, which increases the profitability of that side of the farming enterprise.

Crops are grown amongst the grasses.

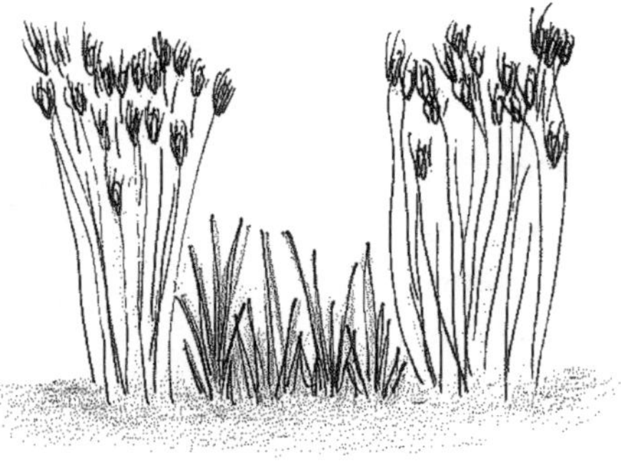

In some ways pasture cropping is better suited for a grazier moving from a pasture system and not that suited for a farmer moving from a conventional cropping system.

Cereal crops can be sown into summer (C4) and winter (C3) perennial native grasses (see chapter 4 for a short explanation about C3 and C4 plants). When a winter cereal crop is grown in a summer perennial pasture, there are distinct growing periods for each species.

Stock can graze right up to sowing and again after the crop is removed, and there is no need for a summer fallow. Because it is not always desirable to put in a crop, you could sow a crop every five years, for example, with four years for pasture and soil recovery.

Whereas conventional cropping methods sprayed all weeds and other plants all of the time, pasture cropping controlled weeds by selective use of herbicides, by livestock management during grazing, and the production of plant material litter to smother them.

It is essential to have complete ground cover at all times, and this results in large increases in plant biomass and soil carbon. However, if weeds are not reduced by some method, then production of both crop and grass is severely reduced.

There is flexibility in what is grown. A mix of shallow and deep-rooted plants will access different soil-water resources, reducing competition and generally increasing biomass production. When the soil-water levels are lowered, there is less risk of waterlogging or dryland salinity occurring.

» DID YOU KNOW?

All soil-based ecosystems have particular proportions of bacteria, fungi, protozoa and other microorganisms, and different levels of complexity within each group of organisms. The food webs in soils are very complex.

Agricultural soils, grasslands and vegetable gardens usually have bacteria-dominated food webs. There are lots more bacteria than fungi, and different types of organisms that feed on the bacteria. In forests and undisturbed bush areas, fungi dominate and the microorganism mix changes accordingly. The biomass ratio of fungi to bacteria in a deciduous forest could be 100:1 and up to 1000:1 in a coniferous forest.

When we change our soils, the balance between bacteria and fungi can change too. For example, if farmers change to no-till methods the ratio of fungi to bacteria increases. Any changes may affect the cycling of nutrients, soil-water interactions, the biodiversity of soil organisms, the overall productivity of the soil, and the filtering and buffering ability of soils to the additions of organic or inorganic materials.

When soil conditions become alkaline, bacteria dominate. This favors weeds, grasses and vegetables. Conversely, fungi dominate when conditions become acidic and this favors shrubs, trees and canes (raspberries and so on). All of this also depends on the soil composition. I have a bacteria-dominated soil, but the soil is acidic as it is derived from laterite containing iron oxides, which is naturally an acidic mineral.

When we use lots of green plant material (high in nitrogen) in our garden, say as mulch, we encourage bacteria; when woody materials and straw (high in carbon) are used then fungi proliferate.

Keyline

Permaculture's main focus has traditionally been as small-scale domestic systems, whereby people live more sustainably. While the principles and practices can be applied to rural properties, there has been little uptake in country areas.

Permaculture practitioners seem to have adopted variations and combinations of other branded methods such as Holistic Management, Polyface and regenerative agriculture.

When rain falls there are only three options: it enters the ground, it runs off the land and enters streams and then rivers (and usually ends up in oceans), or it is allowed to be captured and stored.

Groundwater recharge is a worthy outcome as at least water is stored in the soil (although maybe a long way down), but having water leave the property and head downstream is not ideal. Runoff often causes erosion and loss of top soil into waterways too.

The third option, the capture, storage and use of water, has always been a main consideration for permaculture designed properties.

The movement embraced the ideas of PA Yeomans, who developed the keyline plan in the mid-1950s.

The aim of the keyline plan was to increase soil fertility, improve soil structure and to hold water in the landscape.

The plan involved the identification of keylines in the landscape, new cultivation techniques, better water conservation and irrigation practices, and ways to subdivide farms — the result was an integration of a host of strategies and techniques with the outcome of increased farm productivity.

The keyline plan revolved around the keypoint and the keyline. The keypoint is a particular location in the primary valley and is found where, moving downwards, the steep part of the slope flattens in the center line of the valley. This is where water collects and starts to become the head of a stream.

The keyline is a particular contour line that passes through the keypoint. Either side of the keyline, distinct changes are noticed in the slope of the ground and the spacing of consecutive contour lines on the ridges compared to those in the valley. It identifies the change in slope — from initially being steep to where it flattens out.

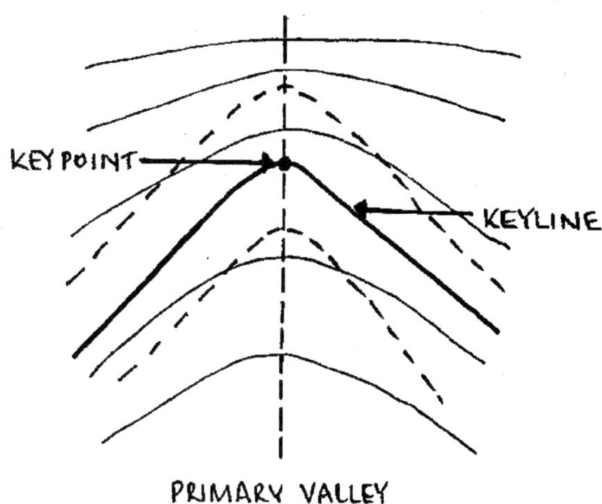

Keypoint and keyline in a primary valley with some cultivation lines shown (dashed lines).

Keylines extend from the beginning of one ridge to the beginning of another across the valley. A keyline ends when the contour changes direction at the beginning of a ridge (this is also the steepest part of the ridge).

Once the keypoint and keyline are located and marked out, subsoil cultivation (minimum tillage) occurs parallel to this contour.

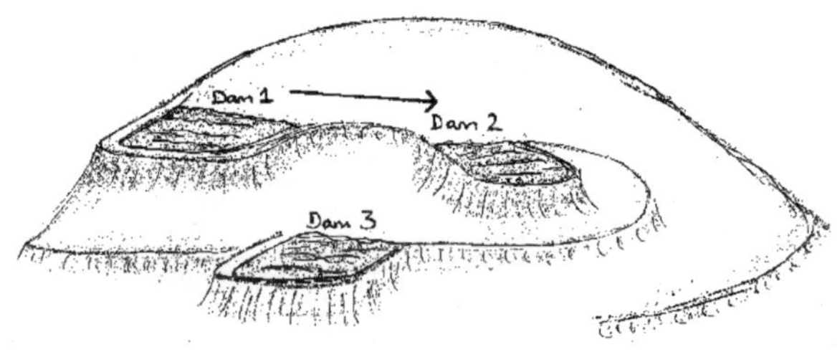

Water can be stored in dams and moved downwards.

This forces runoff and shallow seepage water to collect in the cultivation channels and flow towards the ridges, thus keeping water higher in the landscape for a longer time (rather than letting water just drain down the slope towards a creek).

These cultivated channels above and below the keyline become off-the-contour drains, intercepting water as it moves downwards and redirecting it outwards towards ridges. Ridge cultivation is a little different. The lowest contour on the ridge is marked out and cultivation occurs parallel to and above this. The added benefit is a reduction in erosion and loss of topsoil.

Most of our fragile landscapes (what Allan Savory calls brittle landscapes) need hydration, not draining. Engineers like making drains and moving water away as fast as they can.

Ecologists (and permaculturalists) like to keep water where it falls and hold it there, preventing ridges and hills from drying out too quickly.

The soil acts like a huge sponge and stores water within the soil matrix. Where there is water there is life, and the soil life emerges to begin all those elemental cycles to make nutrients available to plants and so it goes on.

The water slowly begins its regenerative movement through the landscape. Too many people still view water as a plumbing problem and not the asset it really is.

By building dams higher up the slope, water is able to be collected and used for stock and irrigation.

Overflow, either naturally as the dam was full or by through-the-wall pipes, moves by gravity into irrigation channels or from one dam to another down the slope, often around hills into another valley. There can be a mosaic of dams and drains across the landscape.

Yeomans also developed the Keyline Scale of Permanence, which was a sequential list of factors that ranged from permanent in a landscape and couldn't be changed (climate, landforms) to those that could easily be changed on a farm (such as fences, planted trees and soil). This helped farmers plan where to place dams, roads, buildings and tree crops.

The Yeomans family are still involved in developing and enhancing soil fertility and cultivating plows and equipment that are used for this type of work.

The Yeomans Keyline Plow has fixed tines or shanks that slice through the subsoil and break up the hardpan below, all without turning the soil over like most modern plows.

Initially you might set the points in the tines at about 6 in below the surface and then progressively make deeper cuts each year and end up about 15 in below the surface.

Dragging the digging points through the soil enables water, air, seed and manures to penetrate the soil and provide the right materials and medium for soil life to proliferate.

Darren Doherty and Lisa Heenan have expanded the Yeomans Scale of Permanence into the Regrarians Platform (from *Regenerative Agrarian* — an agricultural society). They have tweaked the eight themes in Yeomans scale and added two more — economy and energy.

Whereas Yeomans scale was centered on agriculture, Doherty and Heenan have included social and economic themes. This allows farmers to explore ways in which they can value-add to their products or seek new markets, and develop greater community resilience.

As the main role of plants is to produce food (and other essential organic substances) and oxygen through photosynthesis, the Regrarian Platform also addresses the importance of energy to all life and systems.

Regenerative agriculture

Organic farmers work to improve the soil and our local environment, but those promoting the concept of regenerative agriculture extend this ideology by also advocating the regeneration of community, our health and wellbeing, and the broader biosphere.

Like most organic farming methods and philosophies, it has always been more than the "agriculture part" of farming. But we also need to keep this in context: agriculture is still the backbone of society and the foundation for modern civilisation. We all need farmers!

All agricultural farms are businesses, and they exist to make money for the farmers. Various enterprises are undertaken to generate income and cash flow, and these enterprises can change from one year to the next. Farmers need to make a living and they can do so by incorporating a range of regenerative and restorative techniques.

Regenerative agriculture is an umbrella term for any agricultural practices that lead to building more carbon in soils and thus healthier soils, increasing biodiversity and hydrology in the farming landscape, and still making all of this economically viable and profitable.

RegenAG®, an Australian-based family-owned enterprise, has taken the concept of regenerative agriculture to the next level, by promoting "beyond sustainable" farming. Kym and Georgina Kruse advocate poly-enterprise farming on a single land base, which addresses a spectrum of land-use pressures while ensuring biodiversity and resilience, as well as providing consultancy and training in several regenerative methodologies so that our landscapes and communities can be transformed.

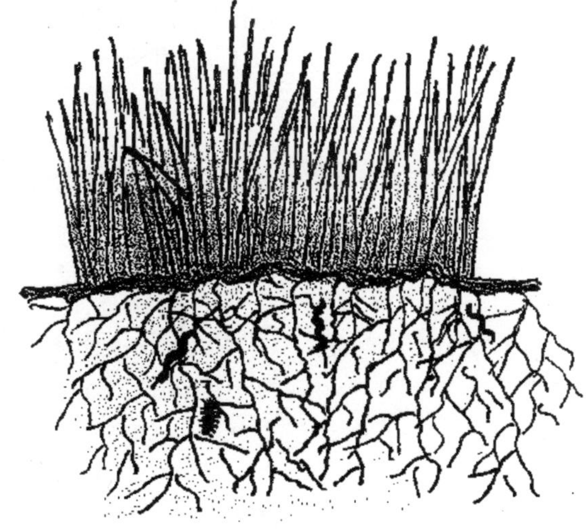

Regenerative agricultural practices increase organic matter (and life) in soils.

Regenerative agriculture embraces aspects of Holistic Management, Polyface farming, keyline, pasture cropping and MasHumus, a biological farming method from Latin America that makes biofertilizer from manures.

Historically, regenerative agriculture was seen as management practices within organic farming, but these principles and techniques can be applied to a wider variety of farming enterprises. There is certainly increasing scientific evidence that these techniques do work, and there are marked improvements in soil and nutrition levels in plants and animals.

Water capture

Keyline is one of the strategies used to harvest and then store water in a rural landscape. Keyline drains are mainly designed to move water away from the valley and towards the ridges.

Other types of drains are employed to capture runoff and move water into dams. Some of these are briefly discussed below.

Swales

Swales are channels cut on the contour. A small bank on the downward side (a berm) holds the water to enable it to soak into the ground. A swale doesn't allow water to flow away like a drain, and is constructed on permeable soil so water percolates downwards by gravity.

As water is kept at that point, trees are often planted to utilize the water. On a farm you could plant fodder species that would both feed and shelter stock.

Like all drains and catchment channels you need to be mindful of slope, expected volume of water to be held and soil stability. A swale made in sand may give way as the soil is not held together that well. If it is expected to hold too much water and the slope is steep then the whole drain wall might wash away.

Before any drains are cut make sure you calculate how much water could be expected in a good downpour over that catchment.

Diversion drains

As the name implies, diversion drains divert water from one place to another. Typically they drain water towards a dam or a garden area. Diversion drains are often called contour drains, but this is confusing as water is moved off the contour. They are also called interceptor drains as they intercept runoff water and permit it to either seep into the ground or drain away.

Water doesn't need much slope to move. A fall of 1:200 (a one foot fall over two hundred feet distance) is enough on larger properties to move the water but ensure minimum erosion of the drain.

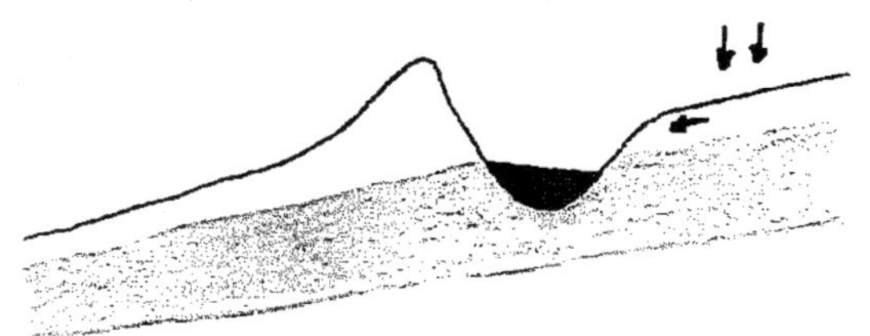

A swale is a ditch on the contour.

Reverse-bank interceptor

A reverse-bank interceptor drain can be used in well-drained or sandy soils. It has the berm or bund on the upward side. As runoff water moves towards the drain it has to filter through the soil bund first. This cleans the water, traps animal manures, leaf and plant materials, and thus prevents polluted water from entering the drain. Cleaner water is then carried to a dam.

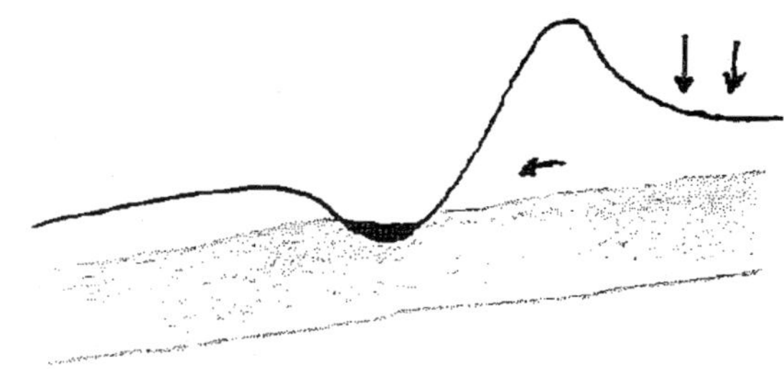

A reverse-bank drain allows a cleaner water to enter the drain.

WISALTS drain

Another type of interceptor drain is the Whittington Interceptor Sustainable Agricultural Land Treatment System, often abbreviated to WISALTS and colloquially known as "Salt Affected Land."

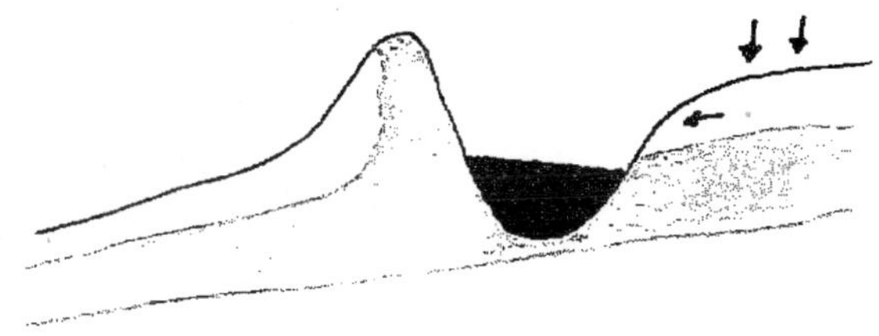

A WISALTS drain cuts into the clay layer and the drain is sealed.

These drains are deep channels that are always cut into the subsurface clay, and then this excavated clay is used to line the drain. The compacted clay layer prevents seepage into the ground and therefore carries all of the salt-laden water away. This has the effect of drying out paddocks and reducing waterlogging problems.

If these drains are not lined with clay or some other liner then they may leak salt water back into the landscape and not perform as they should.

Other aspects of farming

Besides the soil and water management issues we have discussed over the last few pages, farming is also about what we grow. Other chapters deal with fodder plants for farm animals, soil amendments, green manure and cover crops, fruit and nut trees, herbs and vegetables, and other useful plants.

What is important is that we consider farming our land holistically. We make changes only after we endeavor to heal and repair the land, treat all animals humanely and grow our food crops organically.

One of the biggest hurdles for farmers is overcoming fear — will the changes to their farming practices and the associated costs for this be viable? Will they survive and prosper or will the bank foreclose on the property?

To be successful in any business, and that includes farming, you need to develop a business plan. This helps you identify opportunities, evaluate what you have been doing and what needs to change, set goals and future directions, and allow for the ongoing update of your plan as circumstances change. It is basically a road map that lists your aims, ways to achieve these objectives and a working framework to enable change to occur when this becomes necessary.

The farmer's business plan should include the consideration of the possible impact of their activities and enterprises on the environment, and essentially address the triple bottom line: measuring their success through financial, social and environmental outcomes.

The business plan should also include a marketing plan. Farmers who want to diversify their farming enterprises need to identify new markets for their produce, develop a range of strategies to ensure profitable returns and make sure that they can communicate and promote their products to consumers.

Organic food demands a premium price in the marketplace.

Organic food has always been a niche market, but the sales of chemical-free food are increasing, and the demand for pasture-raised meat and dairy products has risen sharply in recent years. So there are lots of opportunities for farmers to adjust their growing methods to gain greater sales at a premium price, and win wider support from consumers.

Modern, progressive farmers are using innovative ways to sell their products. Nowadays, you will find farmers selling at farmers' markets, supplying food to community supported agriculture (CSA) schemes, developing "value-added" products (jams, chutneys, cider), conducting farm tours to keen city dwellers who want to know where their food comes from, using the internet to sell and market their produce (through an online shop), engaging with others in cooperatives, and having their own "farm store" where they sell their products and some from their farming neighbors to an increasing number of willing buyers.

Farm tourism (agritourism) is on the rise, and more people are staying at "farm stays" to experience some sort of country life, so farmers who want even more income-generating ways have to put themselves "on the map" through informative websites, email newsletters, links and associations with both tourism and local community groups, and by offering some "point of difference" that attracts customers.

Community supported agriculture — food boxes

Many people are becoming more conscious about health and food quality. Farmers have opportunities to educate others and to provide valuable information about their products. For example, grass-fed meat generally requires a lower cooking temperature and a shorter cooking time, so "low and slow" is the message to consumers.

Agritourism is one way to connect farmers with consumers.

Like most businesses, it's about developing relationships between people — between the growers and their customers.

Once these friendships develop, then your customers become your greatest asset, as word-of-mouth promotion of your produce will bring even far greater returns.

Whatever you decide to undertake, make sure you do your research. Good old-fashioned marketing and research is crucial when new products are introduced, and poor planning and poor research would be the cause of most failures in business.

Some farmers may be great at growing things but not so good at the "business" side — marketing, budgeting, promoting and selling.

Farmers' markets connect farmers to consumers.

If you have good ideas, seek the services of a professional and you may find the expense may only be a fraction of the income it generates. Like most things in business, you have to spend money to make money.

STRATEGIES FOR SMALL SPACES 12

Many people live in apartments, units and small villas, close to major cities. It makes sense to live close to where you work, to reduce travel time, utilize public transport and generally use less energy to live.

High-density or inner-city living has its benefits and its drawbacks. Everyone needs to be comfortable, have enough space, and just enough "furniture and belongings" to make this possible.

Growing your own food in a dense urban environment can be difficult. Some people may only have a balcony to garden on, others have no such luxury. Whatever your situation, there are many strategies to better utilize the small spaces you have in your home and in the garden.

Inside the house

There are some general principles for being comfortable and content in our homes, and how we can change our situation to ensure this happens. How we organize our homes is best described by the acronym $S^2M^2D^2$. These letters stand for storage, space, monochromatic, multifunction, de-clutter and design. Let's examine each in turn.

S — Storage

Small spaces and rooms seem full even with just a few pieces of furniture. Store small things, seasonal clothes, books you want to keep and some of your music collection to make the room tidy and portray the illusion of emptiness. Built-in and freestanding cupboards, shelving and stackable boxes (with lids) all enable better vertical and horizontal storage. Hooks, pegboards and racks enable objects and clothes to be hung and kept off the floor.

Shelving, cupboards and racks are used for storage.

S — Space

Concentrate on creating negative space — that area of emptiness around furniture. Chairs and tables with legs (with space underneath), mirrors and glass all amplify light and make the room seem spacious. You can look through objects to see other furniture, rather than seeing a solid wall.

Murphy (foldaway) bed

M — Monochromatic

Walls that have the same color (but maybe small changes in tone or color shade) create the illusion of space. You could probably imagine that if there was distinct contrast, such as between black and white, then the eye segregates and defines objects. When a room has shades of a similar color then the eye moves through the room uninterrupted.

Adjustable shelves give you flexibility in storage options.

M — Multifunction

Some rooms can lend themselves to being a play room, music room, guest bedroom or office. Having a fold-up bed in the wall (Murphy bed) is far better than having a bed taking up precious space, set up on the odd chance someone might sleep over.

Multipurpose furniture also allows flexibility in how the room is filled or used. For example, a daybed enables guests to stay over, adjustable shelving enables different-sized books or objects to be stored as required, and modular furniture pieces can be re-arranged or stacked.

De-clutter — put things away

D — De-clutter

Some people are hoarders. When space is at a premium then a room quickly fills with things, many of which are seldom used. Here are a few simple approaches you can adopt.

Firstly, if you haven't used an object after about three years, then maybe it's time for it to go. I store offcuts of materials, boxes of door handles and paint tins that only have a small amount of paint in them, all in the belief that one day these will come in handy. The truth is I've had these things for years, the paint tins have dried out, I never have had to replace a door handle and the offcuts are never the right size or type of material I need. Give useful things away to others or to the thrift shop. The rest can be advertized on freecycle schemes or LETS networks and recycled as much as possible, with the least amount going in the garbage.

Obviously, the time between clean-outs will vary depending on the item. You may recycle clothes after two years but keep a tent for five years or even longer.

Secondly, adopt the principle "something in, something out". When you bring some roadside collection piece home that you simply must have, decide which object or furniture piece it replaces. If something comes into the house, then something leaves the house. Your home should be about life not stuff.

Thirdly, take the approach that if something doesn't give you joy then maybe it should go. When you look at an object it has to make you happy — for whatever reason. If an object doesn't bring a smile to your face or you are not content looking at it, then move it on.

D — Design

Permaculture is about design, and this doesn't just mean gardens. Houses, rooms, layouts of rooms, placement of furniture, balcony gardens and storage systems can be designed to be both functional and aesthetic. Small spaces pose particular problems, mainly because you want the most of everything in the least amount of area. Clever design helps you achieve this.

Applying permaculture to small spaces

We can apply the concepts outlined in chapter 3 about designing gardens to the homes in which we live. For example, whilst "zones" are used to place elements in a garden design, we can also adopt a similar approach inside our homes.

Zone 1 (in a garden) is close to the house and we visit this area quite often. Similarly, the inside zone 1 is where you spend most of your time. The bedroom doesn't count — we do spend a lot of time there but we are asleep. So, do you spend the majority of time sitting at the dining table, on a lounge chair, in your office (at a desk and computer) or cooking in the kitchen?

Zone 1 things are those that you use all the time, so they are often at arm's length as you wander about the house. Zone 1 doesn't have to be one spot but rather a pathway through the house.

Generally things have to be in your face — to see them, to use them and to look after them. So you might place jars soaking in an ice-cream container to remove their labels at the kitchen sink or windowsill so that you see them and clean them, ready for the jam you make the next day.

You are looking for a balance between chaos and calm, between clutter and clarity. We don't need a kitchen bench full of yogurt, kefir and kombucha cultures, just enough to see when we need to act and give them some of our attention.

It's about being organized and efficient (and energy efficient). A small table near the front door can hold your house and car keys, a water bottle, a hat, shopping bags and anything else that you need as you leave the house, so you don't need to panic about forgetting something every time you make a journey — it's all there ready to take.

While zone 1 is not for storage, zones 2 and 3 are. High-use things are placed in zone 2, and low-use things in zone 3. For example, a vacuum cleaner is kept at the front of a cupboard (zone 2) but winter clothes can be stored in a sealed container and placed behind the vacuum or even in a roof space (zone 3, make yourself a mini-attic).

Unfortunately, humans are reluctant to try new things. We tend to want to play it "safe," to engage in familiar territory, to fall back on what we know and what we are comfortable with. So changing our habits can be difficult. People may have the best intentions, but change is hard to evoke and maintain. Change doesn't happen by sheer willpower. Willpower is a highly inefficient source of energy, so we can become complacent. By this I mean that willpower isn't really power (or energy) at all, so we often kid ourselves in thinking that we can do things whenever we want. What happens, of course, is that we stumble and put things off, so changes may start but never finish. We need ongoing motivation and positive reinforcement to continue on a path of change.

Outside the house

There are two components of external house features. Firstly, there is the extension of the internal, and this generally includes furniture, entertainment area and storage. Secondly, there is the garden aspect where homeowners are endeavoring to grow plants, some of which may be edible and some more aesthetic.

External living areas

If you only have a balcony, then place a small table and chairs out there to enjoy the sounds of your environment and just focus on food crops. Gardening is an enjoyable hobby and growing plants and seeing them mature is rewarding in itself. Harvesting and eating them afterwards is a bonus.

If you have a little more room then typically you might have a small shed where you can keep the barbecue, a few tools and items you occasionally use such as tents, bags, beach umbrellas and fold-up chairs. Unclutter the house by storing outside in a weatherproof shed or enclosure.

External garden areas

There are many ways you can utilize a small space to grow plants, and there are many techniques to enable you to grow plants densely. The basic premise is: no space is too small for a garden. Here are some ideas to help you with the design of your intensive garden oasis. None of these strategies are specifically for small spaces and they can be applied in all gardens, but they can work well in small, challenging gardens.

Stacking

Plants with different growth rates and habits can be planted very close to each other. This is a type of intense gardening to create a vegetable polyculture. So fast-growing greens (e.g. radishes and lettuces) are planted alongside slow-growing brassicas (e.g. broccoli and cabbage) and onions. The radishes mature early, followed by lettuces. These can be harvested in turn, whilst the other food crops continue to grow. Brassicas are picked next and then finally the onions (as these mature last).

Stacking — closely-packed plants that may also mature at different times

You might have a dwarf fruit tree in a pot and you could easily plant a range of vegetables and herbs, all with different heights and spreads, as understory. Stack this altogether but ensure each has adequate sunlight and sufficient space as they grow.

Vertical gardening using a trellis...

... and using pots

The secret is to harvest some plants when they are young (as a method to thin crops), not when they are mature. You also need a balance of root and leaf crops. Avoid too many of the larger fruits such as tomatoes, as these will shade out others. Generally remove a whole plant to thin, as this creates gaps to place more seed and allows the growth and development of nearby plants to occur.

Vertical gardening

This is mainly used for vine type crops such as peas, beans, melons and even tomatoes. Trellises, various wooden or steel frames, rows of wires fixed to a wall and wire mesh can all be used as the support structure for spreading plants.

Container (pot) gardening

Many plants, including fruit trees, can be grown in large pots. These can be moved around a balcony or small courtyard to catch the sun as required. Vegetables, flowers and companion plants can be grown under the fruit tree and picked or clipped when needed.

Gardening with pots

Make sure that the pot and mature plant will not be too heavy and awkward to move. Furthermore, when plants are watered the weight of the soil increases so don't put any large pots on balconies that you suspect may not be able to support that weight. You might be able to place large pots on a platform or flatbed trolley with caster wheels to enable it to be pushed around from one space to another without injuring your back.

Container gardening

Picking greens can be grown in foam boxes or similar containers, as long as there are holes in the bottom to enable drainage of excess water. Don't forget to top up the pot or garden bed with new soil each season and to fertilize regularly.

Window boxes

These are containers, usually long, rectangular and only about 12 in deep. These containers are mounted in front of, and below, a window outside the house, so that vegetation and color (window boxes are also called flower boxes) can be seen as you look out the window. Window boxes are mini gardens that enable you to bring the outside in. They also make a decorative feature and brighten an otherwise drab exterior wall.

Window box

Succession planting

There are a few variations of this technique. You could plant a number of varieties of say tomatoes where each matures at different times (early, mid and late varieties), so the availability of tomatoes is extended over a longer time.

Secondly, you can start with one crop and then before it is finished plant the same type again (or plant seeds one month apart over a few months).

This "succession in time" idea was developed by Masanobu Fukuoka in Japan and is now practiced throughout the world. Suitable plants to use are lettuce, spinach and radish, which all have long growing seasons but can be harvested within weeks in some cases (or at least some leaves can).

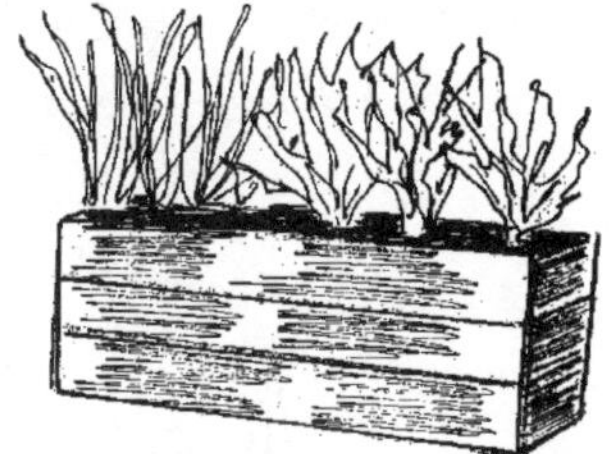

A raised garden bed

Raised garden beds

These can be made in varying heights, so that growth is staggered vertically. Plants in the lower beds shouldn't reduce sunlight access to taller beds, and vice versa. The beds themselves can be just 1 ft wide, although bought steel beds are usually wider than 2 ft as steel crimps when bent too tightly.

Wicking beds may be a useful strategy in some situations. These have a base or container at the bottom to hold water. Water moves up by capillary action to make the soil moist, enabling plants to access moisture as required.

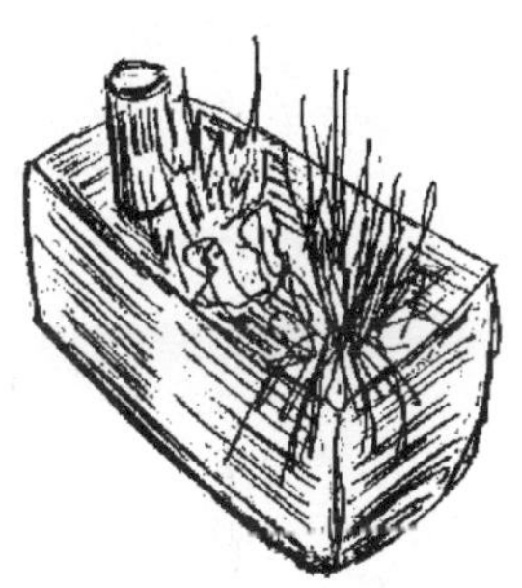

Wicking bed — a self-watering garden bed that is ideal when normal irrigation is not available

Extending the growing season

You can lengthen the growing season of your plants by changing the surrounding microclimate. Typically, this is done by providing protection from excess sun, wind and frost.

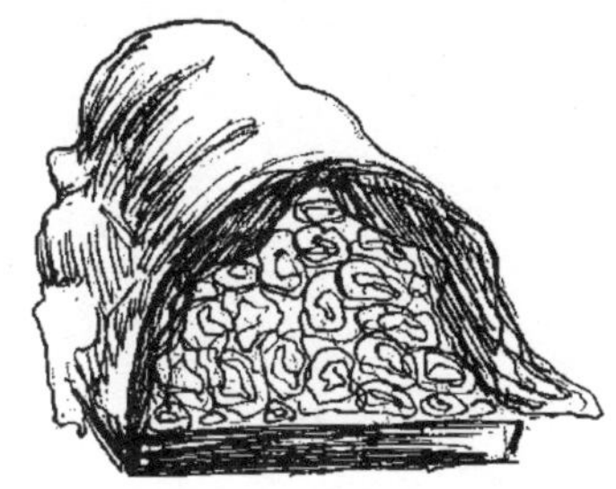

Cloche — a cover over growing plants

Plastic or glass cover for individual plants

You can do this by placing your plants in a cold frame or covering them with clear plastic or some other membrane to trap heat. If your plants are exposed to too much sunlight then placing shadecloth over them reduces their stress.

Exploit the aspect

If your garden space or balcony is in full sun for most of the day, then sun-loving varieties such as tomatoes, rosemary and peppers will thrive.

However, if you have shade for a large part of the day then plant shade-tolerant species such as silverbeet, lettuce and many other "greens." If you have pots or small containers then move these around to exploit the sun, or shelter from it.

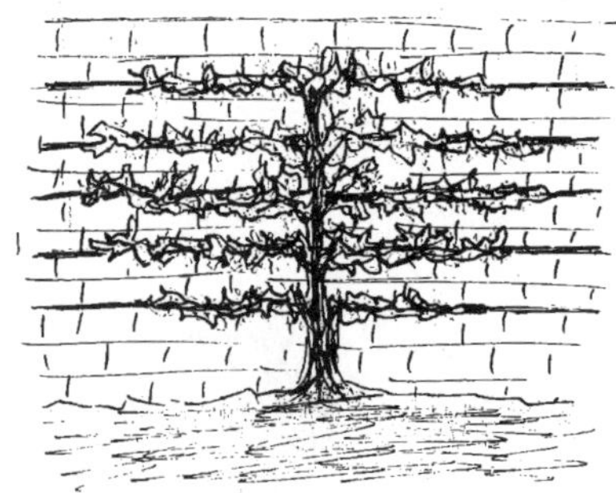

Espalier fruit trees on walls that have a sunny aspect.

Balance of plants

Try a mix of perennials and annuals. Most vegetables are annuals but a few are semi-perennial such as rhubarb, globe artichoke and asparagus. These either naturally die back in winter or are cut back deliberately, regrowing in spring or early the following year.

Endeavor to find out what plants grow well in your area, determine what foods you like to eat or are willing to try and then stick with these.

Don't grow too many of one type as you will just have too many tomatoes — one plant is ample. Variety is the key, so plant only a few of each vegetable type and choose varieties that are hardy and productive.

Block planting

Block planting

Rather than spreading your lettuce seedlings throughout a bed, simply plant them all together in a block. These can then be plucked and thinned as they grow. You do use less space when a group of the same plants are grown like this.

Beetroots, onions and carrots can be planted in blocks, and you can try to see if this technique works for the other vegetables you like to grow. Square foot gardening is a variation of this idea.

Living walls

These are also known as green walls or vertical gardens, and many are hydroponic setups that use little or no soil. Plants are suspended in an inert, loose fiber or spongy-block media. Pots or containers, full of media and plants, are suspended in a frame. Irrigation pipework is interwoven throughout the frame to water the plants. Some of these pot setups are like mini wicking beds and have a small reservoir of water to supply plant roots.

A living wall from recycled pallets.

As an alternative use a recycled plastic or wooden pallet (wood tends to rot though), fix to a wall, line with shadecloth and fill with soil. Where you want to plant, cut the shadecloth away and carefully place your seedling into the soil. This is easy for seedlings, very difficult for mature plants.

Before you start fixing trellises or living walls, make sure you have permission from your landlord.

Hanging baskets

Not only can you grow from the floor upwards, you can hang pots and baskets from the ceiling downwards. Depending on the height you have above you, several baskets can be suspended one under another. Water applied to the top one filters downwards into the next basket.

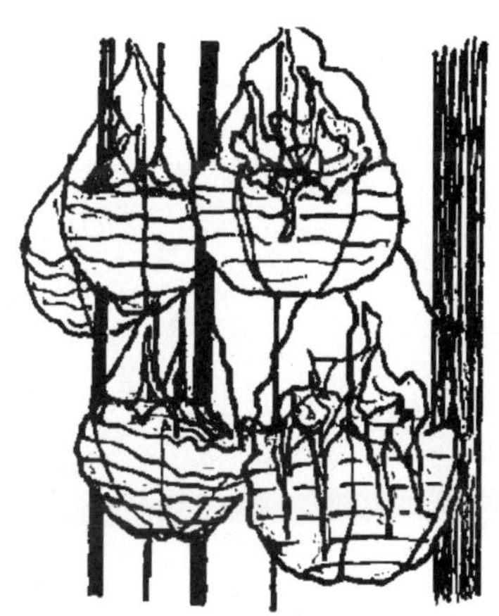

Hanging baskets can contain rampant plants.

Hanging baskets can be traditional fiber-lined wire mesh types or you can use sections of cut-in-half plastic sewer pipe or gutters with the ends capped.

Picking greens such as lettuces, chives and spinach thrive in baskets, while strawberries and mint can be contained and allowed to cascade over the edge. Potted color adds variety and aesthetics (who doesn't like seeing colorful flowers?) so mix ornamentals with food crops. Hanging baskets can dry out quickly in warmer weather so make sure you monitor moisture levels and irrigate or hand-water frequently.

Watering your plants

Various waterwise techniques are outlined in chapter 10. When you only have a small space then large impact sprinklers are not really suitable. Wicking beds are self-watering, but you still need to top them up once a week or so.

Use old gutter or PVC pipes to make vertical hanging baskets.

You may be able to set up a small reticulation system, with a controller and solenoids, if you have some garden area on the ground, but if you are balcony-gardening then hand-watering is probably the only option. Buy a watering can and keep it nearby.

Reticulation systems should incorporate dripline irrigation, and watering should be undertaken meeting local guidelines and enforcing any local restrictions (certain days, times of the day, duration).

It is always a good idea to consider drainage when watering on balconies, especially as water comes out at the bottom of a pot or garden bed and can pool on the balcony. Ensure that excess water can drain away safely and not drip below onto your neighbor's property.

Composting

How to compost organic waste is discussed in detail in chapter 4. Even in a small space you should endeavor to place your food scraps in a compost bin, a bokashi bucket or a worm farm. Materials generated from these systems are ideal to reuse in the garden and will provide fertilizer and mulch for your plants.

Composting leaf material and prunings from plants may be possible but you generally need a large space for this to occur properly. Compost bins can be used to store materials but they only ever produce a cold compost themselves.

Additionally, you may find vermin (including insects, spiders, rats and mice) are attracted to the free food and shelter you are providing for them.

Some closing thoughts

You can adjust and live anywhere in the world provided you are happy in that space. Make it your own. Put your favorite things in view, or move them around occasionally, so you can see and appreciate them.

Change the setting by displaying stored items and storing others you may have tired of. People tend to fill large spaces with more possessions, so living in small spaces can be a great way to reevaluate your "stuff" both on a physical and an emotional level.

Keeping a small space neat and tidy, especially the kitchen and other work areas, is essential as small messes tend to morph into large messes. And whenever and wherever you can, garden and grow food.

HAND TOOLS

13

THIS SECTION ONLY DEALS WITH HAND TOOLS, and no electric or electronic tools and equipment is discussed. Everyone should have access to a reasonable range of tools, especially if they want to develop skills in repairing and modifying structures, and undertake building projects.

Personal protective equipment (PPE) and general work clothing

Even hand tools are dangerous, and can cause damage to our bodies. The use of PPE when undertaking any work, gardening or otherwise, is basic good sense. I am sure we can all remember times when we stubbed our toes, cut our fingers, got dust in our eyes or got sunburned when working outside all day. Here is a list of the types of protective gear we should use and what clothing is most suitable for manual work. All of the following tools and items are listed alphabetically in each section.

1. Mask

A dust mask is mainly used when sawing timber and handling compost and soils. A cartridge gas mask is essential when using volatile chemicals. These often have a fine filter and a carbon filter to absorb substances that smell.

2. Earmuffs or earplugs

Good earmuffs offer better protection, but earplugs are a lot less clumsy and easier to put in your pocket to use when required. They need to be inserted properly to be effective.

3. Gloves

Heavy duty safety and work gloves differ from gardening gloves, where you need some sensitivity to what you are handling and doing.

Cheap polyurethane, cotton-backed gloves are ideal for gardening work. Waterproof gloves are best if you are mixing cement by hand or using chemicals of any sort.

4. Hand gel

We just can't be too careful when it comes to our health. Use a sterilizing hand cleaner or gel (or at least wash your hands with soap) after handling manures, compost, potting mix and any wastewater apparatus, materials and equipment.

5. Hat

You don't need a hard hat unless you are working on a construction site. A simple floppy hat, cap or general-purpose hat is essential for all outside work.

6. Kneeling foam pads

Or individual knee pads for tiling, weeding or carpet laying. Too often we suffer because we are too lazy to get a knee pad and we damage our knees or back.

Keep a few foam kneeling pads in sheds and around the house, and you will wonder why you didn't use these when you started working years ago.

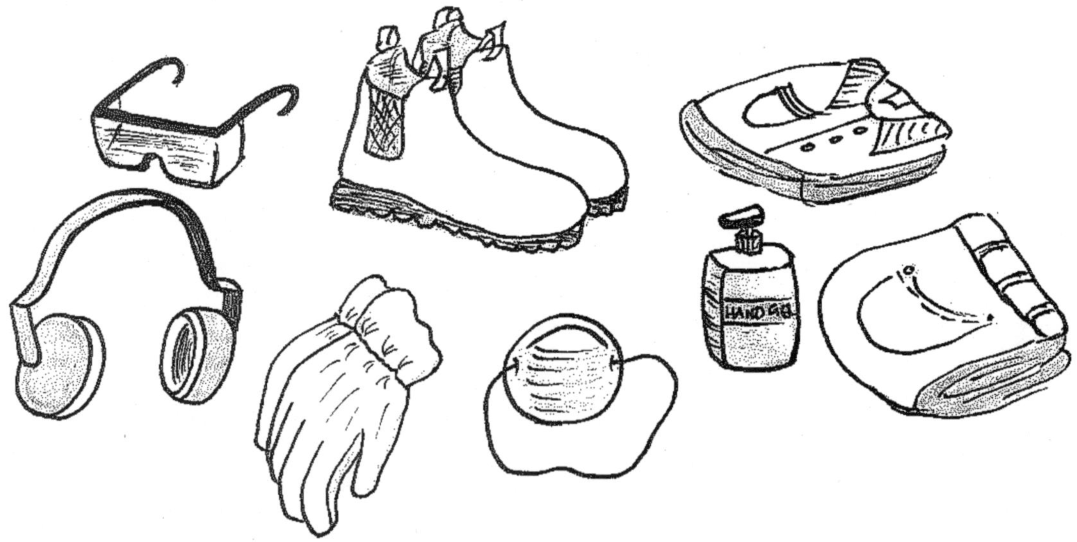

7. Long pants and shirt

Ideally, cotton and loose-fitting clothing should be worn. Many artificial fibers are extremely flammable so such clothing should not be worn if burning off.

8. Sturdy boots

You don't need steel-capped boots for gardening, but something better than sandshoes is a must. Steel or fiberglass capped boots are required on most building sites.

9. Sun- and safety glasses

You can buy one pair of glasses that provide both functions — sunglasses to minimize glare and sun damage to your eyes, and safety glasses that protect the eyes from flying objects — and that includes insects.

10. Sunscreen lotion

Cover up. Many regions in the world are receiving more ultraviolet light and more intense sunlight. Smother your face, arms and any skin part of the body in direct sunlight.

Many people have a medical cabinet or first aid kit in the house, but smaller versions of these are practical in other areas, closer to where you work.

It doesn't take much effort to put a few Band-Aids, tweezers, sling, antibiotic cream, hand lens, small bottle of tea tree oil and headache tablets into a plastic container, and keep it in the car or garden shed or garage. The hand lens, tweezers and tea tree oil is basically the "splinter" kit, as this is probably the most common injury you will sustain.

So far, we haven't mentioned cleaning materials. This includes soaps, rags, solvents — all of which need to be supplied and available as required to wipe surfaces clean, wipe hands, and generally help you to clean up after you complete work.

There will also be conjecture about any of these lists. If you live in woodland or forest areas, a hand ax would be high on the list, in tropical areas then a machete would be invaluable. On the other hand, too many tools and the tool box becomes heavy. Choose tools that you know you will use.

Ten basic hand tools

These are tools that most people regard as essential for home and garden maintenance. Buy reasonable quality tools if you intend to undertake garden or handyperson tasks regularly.

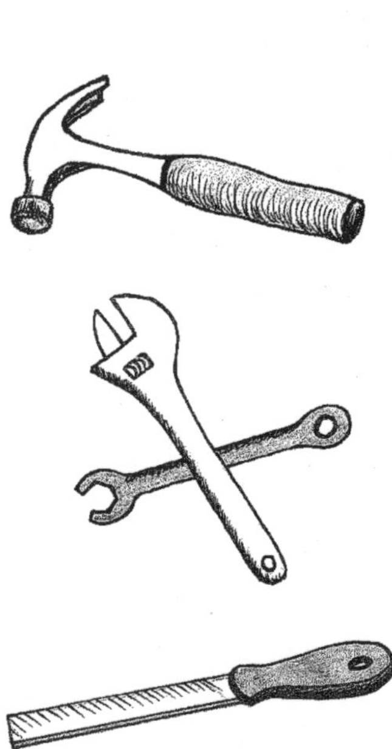

1. Claw hammer

Used for embedding and removing nails in timber. A good-quality hammer will be able to be passed on to your children.

2. Crescent wrench and spanners

Open ended and closed-ring. A crescent is an adjustable spanner, but also a set of spanners for different-sized nuts. A set of imperial and metric spanners are invaluable for any mechanical work.

3. Files

Buy a set of metal and wood files. Priceless tools to clean surfaces, shave wood to enable a window to close, and to trim and de-burr metal pipe and tube.

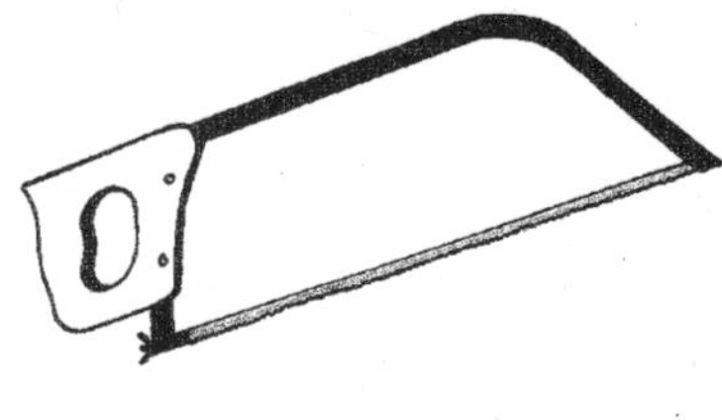

4. Hacksaw

For cutting metal, plastic and wood. Blades have different numbers of cutting teeth, each for a particular material or application.

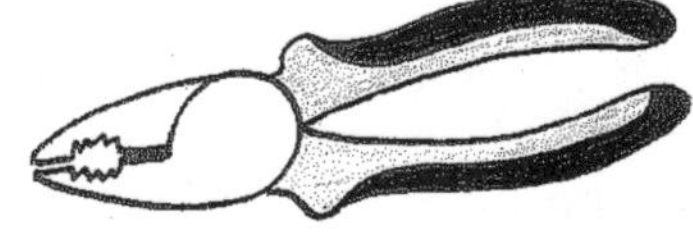

5. Pliers

Useful for twisting wire, grabbing and holding small objects. Long-nose pliers are used for those situations where it is difficult to insert the pliers into a confined space to hold something.

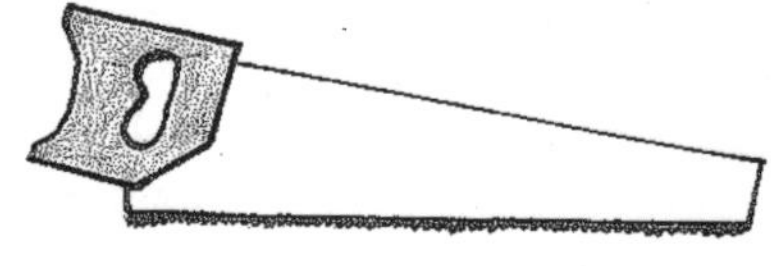

6. Saw

Used for cutting timber or plastic sheeting (and PVC pipes). While you can sharpen the teeth after they become blunt, this is time consuming and besides a special file, you need a teeth-setting device to offset the teeth. It is not uncommon, but in my view wasteful, to literally throw away a saw after each job. You should seek out services of a "saw doctor" to sharpen blades.

I have a few saws, still useable, that I inherited from my grandfather.

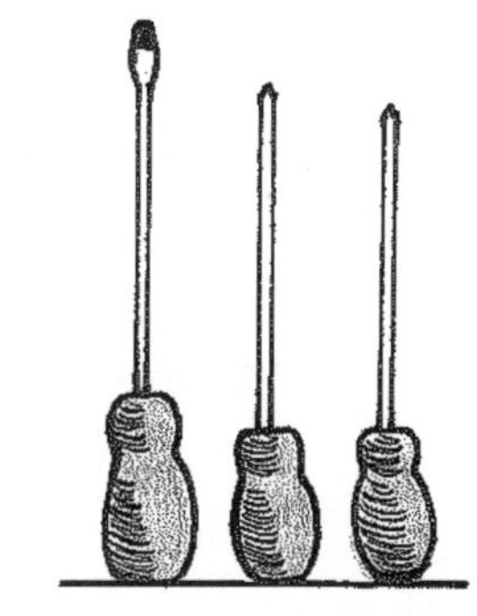

7. Screwdriver set

Flat blade and Phillips head. Get about four or five different-sized blades and Phillips head (cross) screwdrivers. Occasionally you might require a precision screwdriver set. This is a range of small flat and Phillips head screwdrivers for electronic equipment, computers and so on.

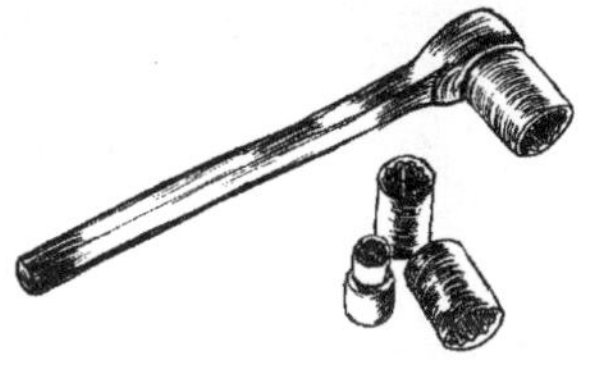

8. Socket set

Invaluable for tightening and loosening nuts and bolts. Most sets contain a fair range of imperial and metric sockets of the most common sizes of nuts.

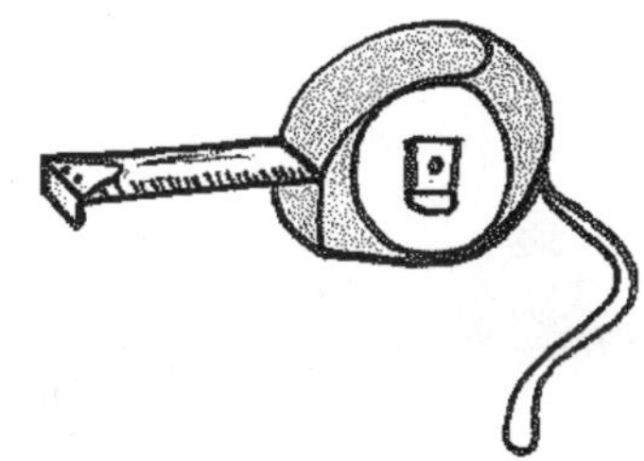

9. Tape measure

A 25–30 ft tape will cover most things at home. Use a 100 ft or 150 ft tape if you are measuring in the yard. You probably need a pencil as well. There are special "builders" pencils but any marker, graphite pencil or chalk stick would suffice.

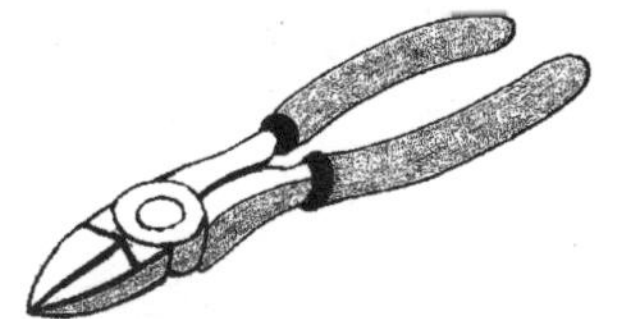

10. Wire cutters

These can cut and trim wire and electrical cable. Will also cut cord and string. If you will be undertaking irrigation work and need to trim solenoid wire then wire strippers are essential. Wire strippers also remove plastic sheathing on electrical cabling.

Handyperson set of ten tools

These are necessary for general maintenance and small building projects. Most people possess a tool kit of some sort, even one for their car, but here is a list of "must-have" useful tools.

1. Chisel

Buy a few different ones for removing and cutting wood, a cold chisel for metal and a bolster to fracture concrete and cement.

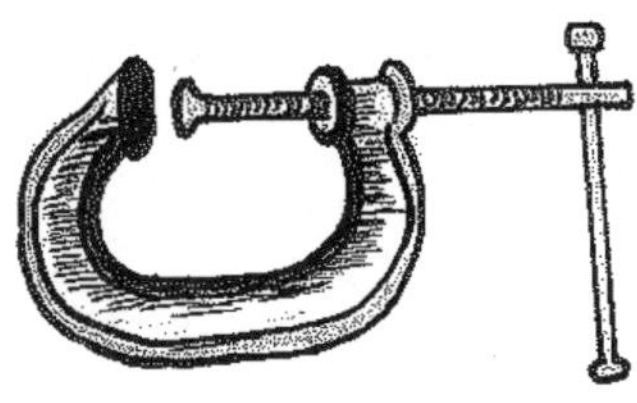

2. Clamps

There are so many different types: sash clamp, G clamp, Bessey clamp and a pipe clamp. These all hold materials and objects while they are being glued, welded and fixed (nailed or screwed).

A cheaper version of the sash clamp is the adjustable pipe clamp. You can make such a clamp almost any size as you screw the head onto a threaded steel pipe and slide the fixing/holding end along the pipe to the desired length.

3. Club hammer

Also known as a lump or mash hammer, a club hammer is used for masonry and concrete work, and belting and bending metal.

4. Hand drill and bits

Yes, you can still buy a hand-powered drill. It's good exercise but a lot slower than an electric drill. Bits are the "drills." It doesn't matter if you have imperial or metric as there are corresponding alternative sizes for each. Drill bits can be sharpened with a grinding wheel or "bit sharpener."

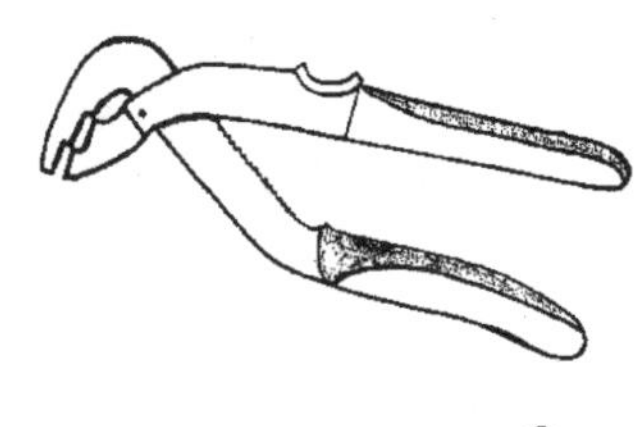

5. Tongue-and-grove pliers

Multifunctional tool for irrigation work, electrical and plumbing jobs, and holding different sized materials.

6. Paint scraper

Flat-bladed tool to remove paint from timber, dried glue and silicone from surfaces, and flaky rust from metal before you weld. Use a metal file to keep sharpened.

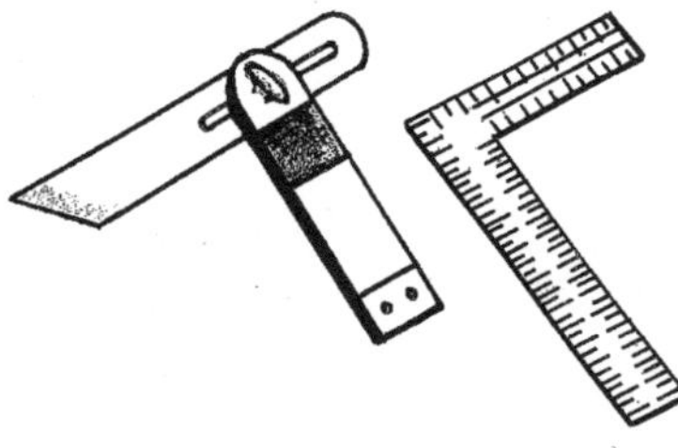

7. Square and adjustable squares

For measuring and marking straight lines and angles (90° and 45°) on materials. A combination square is a common variation and the size of the arm can be adjusted. A bevel is an adjustable square so that any angle can be marked.

8. Spirit level

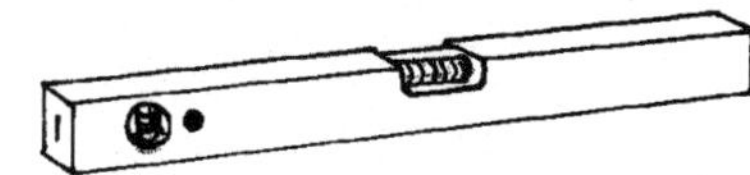

A long 4 ft level is most useful for construction work. These ensure walls and structures are horizontally and vertically level. You can purchase shorter and longer levels, but this size is adequate. A small 12 in level is often used by plumbers to gauge the fall (slope) in pipes and fittings, and it could be useful in tight spaces.

9. Pipe wrench

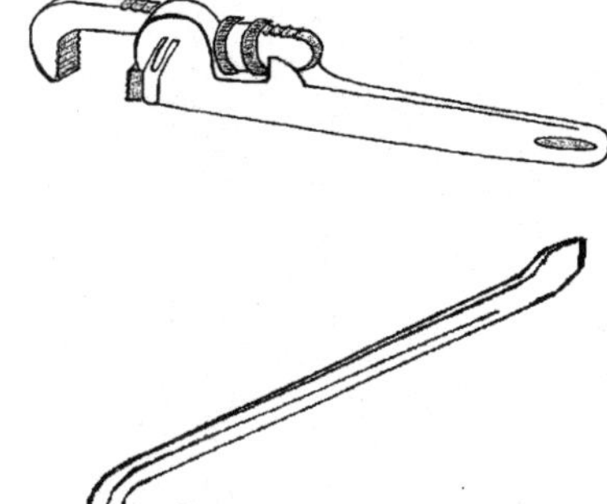

Adjustable wrench so you can unfasten or tighten threaded fittings and pipe. Essential for plumbing and irrigation work.

10. Wrecking bar

Commonly called a "jimmy" or "jemmy" bar. Provides leverage to lift and shift objects, and to remove nails and screws from timber. Crowbars are mentioned in the garden tool section.

Supplementary set of ten tools

Additional to the basic and handyman tool kits, there are some tools ideally suited for building and gardening projects.

1. Allen keys

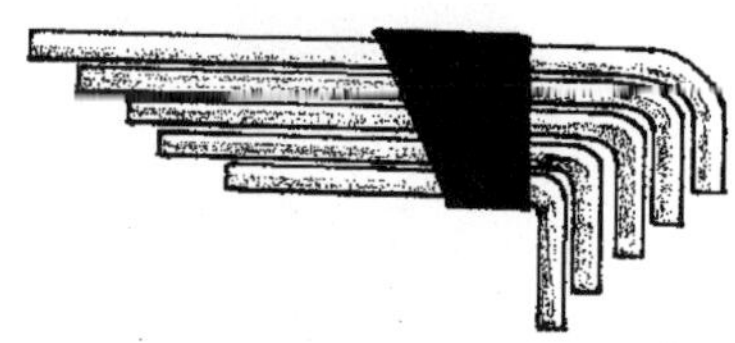

Metric and imperial set. Often used to assemble beds and other furniture that comes in a "kit." Also used to lock drill bits in hole saws and hold components in some electronic equipment.

2. Bolt cutters

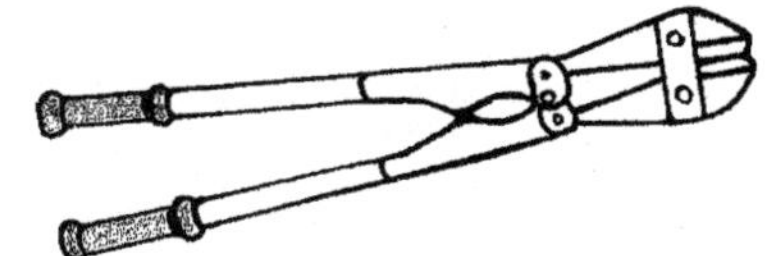

Not used very much but exceptionally handy if you need to remove a padlock because you have lost the key, cut thin steel rods for stakes and concrete mesh sheeting to size, and rusty bolts when you just can't undo the nut.

3. Caulking gun

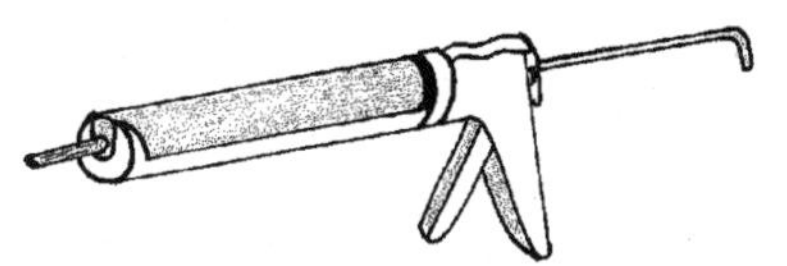

Enables silicone and other glues, fillers and sealants to be squeezed out of the tube onto a surface or into a joint.

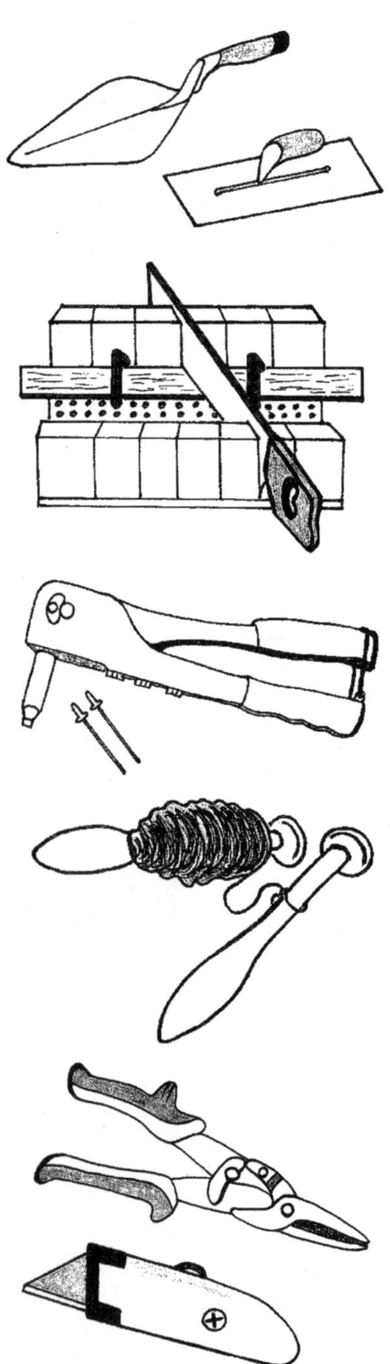

4. Cement trowel

Besides the obvious use for brick and block laying and mixing small amounts of concrete for repairs, cement trowels substitute for a garden trowel to dig small plants and seedlings in and out.

If you want to undertake concrete and cement work then it is useful to obtain a cement float, a "rake" for bricklaying and some formwork.

5. Miter box

A wooden or plastic template to enable accurate cutting of materials at the right angle. Standard precut templates are for 90° and 45° (miter joints for picture frames, flyscreen windows, architraves and skirting).

6. Rivet gun and rivets

Generally used to secure metal to metal, such as a pop in a gutter and the gutter end plates, but also useful to fix plastic and fiberglass sheeting to each other or to other materials.

A multipack of rivets will provide a range of different sizes and most rivet guns come with different nozzles for each rivet size.

7. String line

Usually colored yellow or pink, these fine cords can be stretched (within reason) without breaking. Ideal for measuring out garden areas, building envelopes, fence lines and brick courses.

8. Tin snips

If you think you might use these more than occasionally, get the three different types — right-handed, left-handed and straight. These will enable you to cut sheet metal from any angle or side.

9. Utility knife

There are many versions of this cutting and trimming knife, and most have retractable blades that can be easily replaced. You can cut and size washers for taps, trim excess insect screen from window frames, and cut rope and string to length.

10. Vice grip

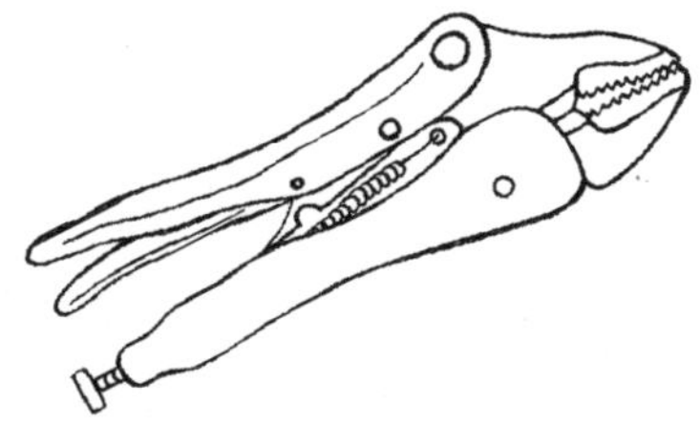

This is an adjustable lock grip. How many times have you wished for a third hand to hold something while you were trying to secure or repair? A vice grip enables you to clamp two objects together while you solder, weld, glue, screw or nail to fix together.

We haven't discussed sharpening tools. This is a skill that can be developed and it is not hard to learn how to sharpen chisels, ax blades, rake teeth, shovel blades, shears, screwdrivers, drill bits, and even saw teeth. Get yourself a few diamond sharpening blades and handheld sharpeners, grinding stones, a few files, some fine machine oil and a grinding wheel, and you will have the opportunity to sharpen almost all tools.

Ten basic garden hand tools

Again, most people who do some gardening will have a basic set of tools, typically stored in a garden shed. You can spend a small fortune buying tools, so just slowly accumulate them, picking up a few bargains from swap-meets and garage sales. Ask family and friends to buy you tools for your birthday.

1. Broom

Cleaning up after a job is part of the job. Sweep soil off paths, spilled mulch off driveways and, upturned, a broom can be used to level small soil areas.

2. Crowbars

These are heavy, straight, solid metal rods with a pointed end or a chisel shaped end to help break up hard ground, concrete or split soft rocks.

3. Garden trowel

Ideal for digging out weeds and planting seedlings, although at times you might want to simply get your hands dirty and feel the soil. For deep-rooted weeds in lawn areas a weeding fork is useful.

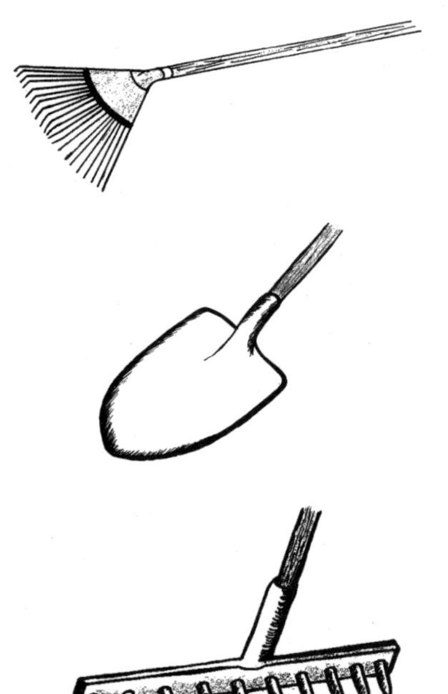

4. Grass rake

Usually a plastic rake (but can be lightweight metal) that scrapes up lawn clippings and weeds in the garden and soil on paths. Not that suitable for heavier weed and plant piles or coarse mulch (you would use a metal rake instead).

5. Long-handled shovel

Shovels can be round-mouthed or straight-mouthed. Long-handled shovels reduce backaches and soreness as you don't bend as much while using them to shift soil, compost or mulch. Shovels have an angle between the handle and the metal blade, whereas spades tend to be straighter.

6. Metal rake

Strong, pointed teeth and a strong rake head enables the easy grasp of objects and garden materials. Some rakes have a spreader bar (plate) on the back to enable you to flatten and level sand.

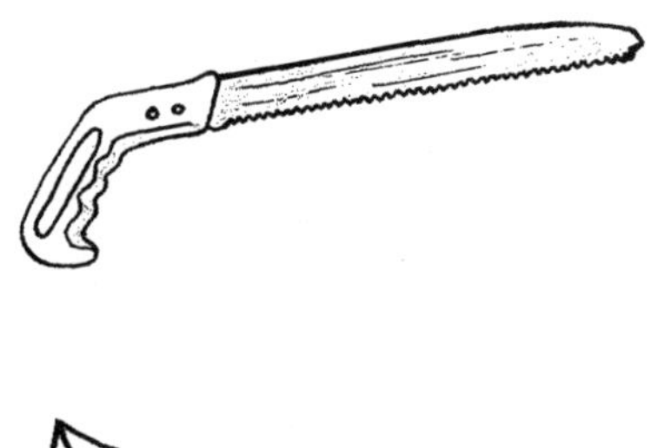

7. Pruning saw

The problem with trees is that they grow. Often they grow too big for us to easily pick fruit or they reach towards overhead power lines, or start to shade everything else out.

Pruning saws have teeth set at a particular shape and size so that soft wood is easily cut. Pruning is an art and requires training and practice. The Japanese style turbocut saws are well worth the cost and new blades can be bought as replacements.

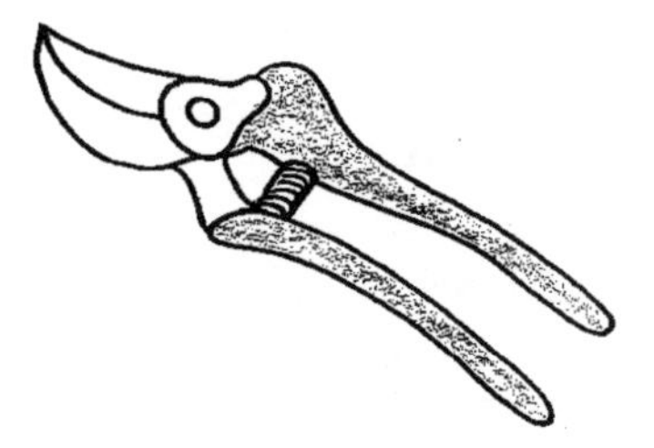

8. Pruning shears

Used to prune and trim plants, and take cuttings for propagation. The biggest problem is neglect — they dull easily and need sharpening on a regular basis. Pruning shears are not that difficult to take apart to replace damaged blades or to sharpen the blade on a fine grinding stone or diamond file. Oil the spring, keep in a holder, and keep out of the weather. Spoil yourself and buy a reasonable quality pair.

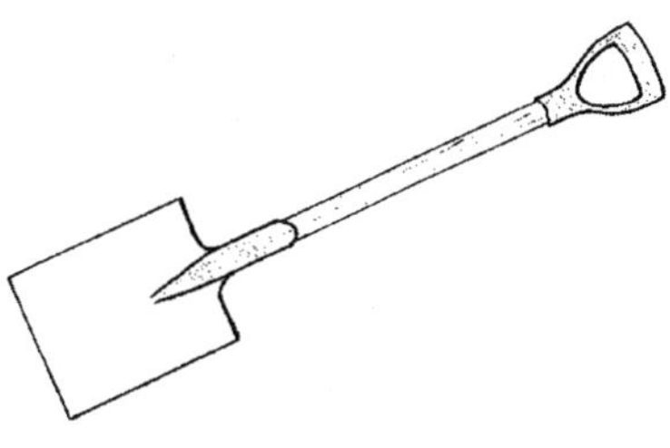

9. Spade

A straight-handled, straight-bladed tool to vertically slice through soil. Enables you to cut lawn so you can remove it or dig holes for planting shrubs.

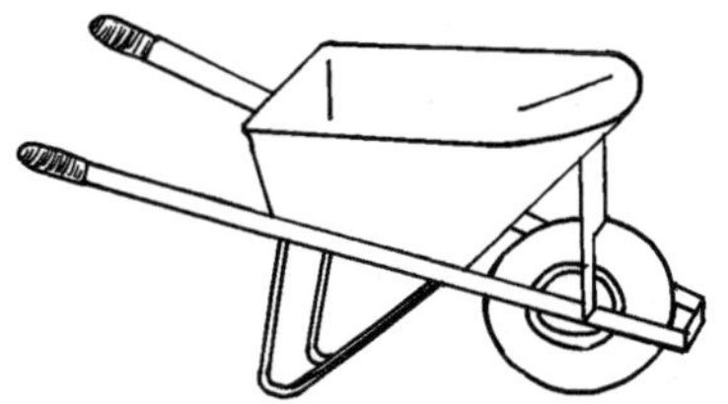

10. Wheelbarrow

Find one that suits your style and lifting ability. Pneumatic tires are generally better than solid rubber tires. Keep the axle greased, tires pumped up and handles and tray clean.

Ten handyperson garden tools

For the serious gardener (and shouldn't we all be) here is a list of some common and some "not-so-common" tools.

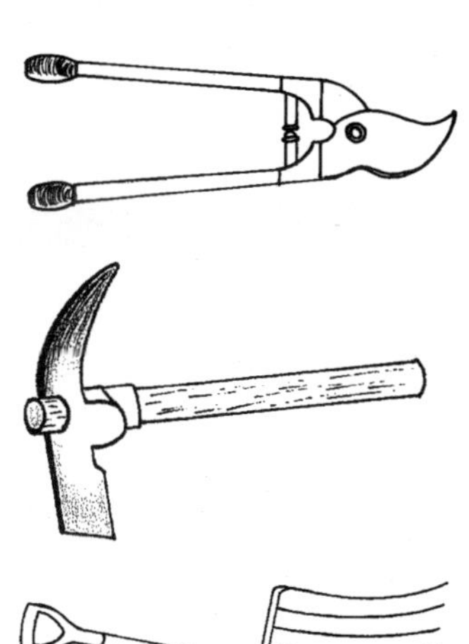

1. Long-handled pruning shears or loppers

Larger diameter branches require a little more leverage and bigger blades to cut through them.

2. Mattock

Like the pelican pick, this tool is useful for digging and breaking up hard soils. Having the mattock (flat blade) on one end and a pick (pointed blade) on the other end is a good combination. Great to remove large weeds!

3. Mulch fork

Much bigger and more expensive version of a pitch fork (a cheaper alternative but not as useful). Great for moving coarse mulch and piles of weeds and lawn clippings.

4. Pelican pick

Underrated digging tool. Enables you to clean out trenches, dig and scoop soil from ditches and score the ground to mark out

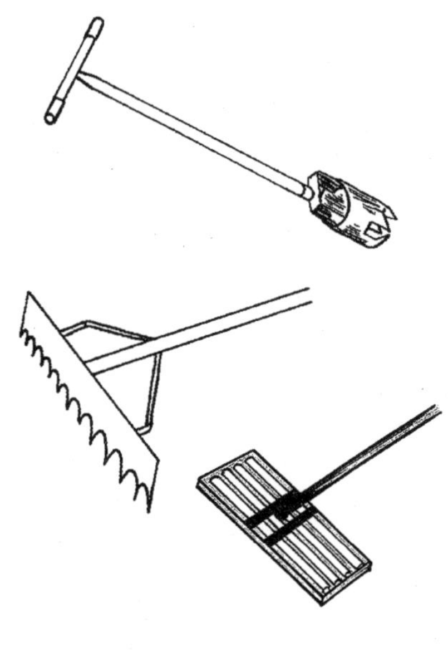

drainage and irrigation lines. Use a pelican pick instead of the normal pick or mattock.

5. Post-hole digger

Not so important in sandy soils, but useful for heavier soils. If you want to plant small shrubs and tubestock, then a 4 in or 6 in post-hole digger is great.

For larger, deeper holes think about using a Dingo or Bobcat with a hydraulic digger.

6. Sand rake

Typically an array of round rods or tubes fixed as one large plate that can swivel with the handle to flatten and level soils.

There are many different tools that are used to flatten sand.

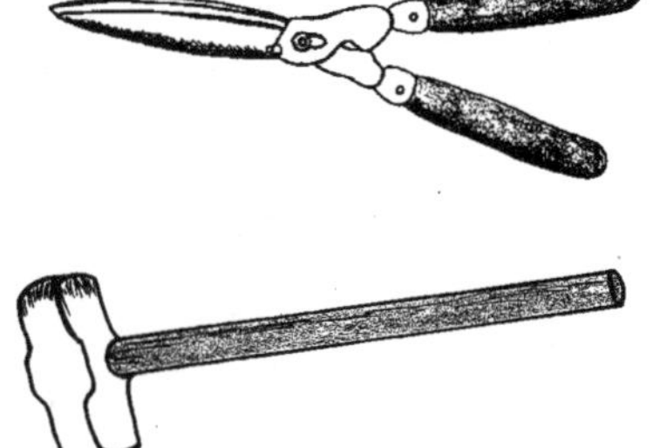

7. Shears or hedge trimmer

Can cut and trim lawn and hedges, and shape plants as required.

8. Sledgehammer

These are identified by the weight of the head. The common sledgehammer is a 8 lb (about 3 kg) hammer. Useful for breaking up concrete, rocks and cement blocks, and driving posts and star pickets into the ground. Lighter versions also exist.

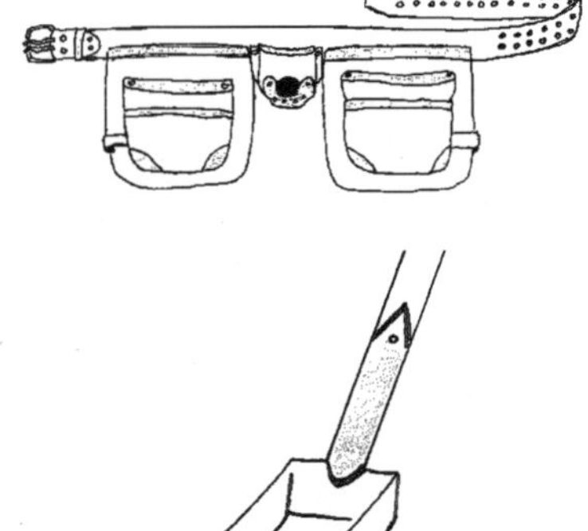

9. Tool belt and nail bag

These vary in shapes and sizes, but even a single pouch with a couple of tool pockets for screwdrivers, hammer and pliers would suffice. Nails, tek screws and bolts are kept in the main pouch (the nail bag).

10. Trenching shovel

Usually only for occasional use, but great for irrigation lines. One of the most-used tools for landscapers, irrigators and plumbers. Sometimes you just need a small amount of soil removed to lay a pipe.

FOODS YOU CAN EASILY MAKE AT HOME

14

PEOPLE HAVE LOST THE ART OF preparing, cooking and value-adding to the fruits and vegetables they produce at home. It all seems too complicated as people prefer not to cook their produce and would rather eat takeaways or just open a packet and microwave it.

Furthermore, when we are stressed (physically or emotionally) we tend to eat "comfort" foods; those typically high in sugar to provide immediate energy. Comfort foods include pastries, pies, cakes, pasta and so on. You can eat these in moderation.

Besides, processing removes fiber and many vitamins and minerals, so raw and minimally-cooked foods are best. No wonder people have real trouble about food — they don't know how to grow it, how to use it, how to store it and how to change it. How do we encourage people to reassess their food habits?

The criteria for the following recipes are that they are easy to follow and the products scrumptious to eat. Not all of the ingredients you will use are homegrown, but certainly some can be.

» **NOTE: CONVENTION USED IN THIS BOOK**

Teaspoon = tsp

Tablespoon = tbsp

Ten simple foods to make

Tasmanian cookbook author Sally Wise was instrumental in helping me put this list together, and the following are small variations of her tried and tested recipes. So here are Sally's top ten (not in the usual cookbook order you might expect, but simply listed alphabetically).

1. Apricot chutney

This is much fruitier than traditional chutneys, which often contain a higher proportion of lemon juice, vinegar and spices that make the garnish a little tart.

Ingredients

3 lb apricots (or peaches or mangoes)
1 lb chopped onions
2 cups vinegar
1.5 lb raw sugar
1 tsp salt
1 tsp mixed spice
1 tsp ground cloves
1 tsp curry powder

Method

1. Chop fruit up. Combine all ingredients in large pot, stirring until sugar dissolves. Gently boil for 1 hour until chutney is thick.
2. Bottle and seal immediately. Store in cool dark cupboard.

» DID YOU KNOW?

Chutneys are made when the ingredients are finer and cooked for a long time (several hours), while relishes are chunkier and cooked for a shorter time. Sauces are finer still (put relish or chutney through a food processor and strain). Most of the sauces and chutneys can be changed to suit your needs, and you can make your own "signature" products. Pickles are produced when foods are preserved in brine (salt solution) or vinegar.

Apricot chutney

2. Apricot tea cake

Apricots are low in fat and cholesterol, high in fiber, a good source of vitamins A and C, but are also high in sugars and have a relatively low amount of most vitamins and minerals (<5% of daily allowances).

Ingredients

1 egg
¾ cup raw sugar
1½ cups flour
2¼ tsp baking powder
¾ tsp salt
½ cup milk
1 lemon
2 oz butter
1.5 lb apricot halves (or use peaches or pears)
Topping: 1 oz butter, 1 tsp ground cinnamon and 3 tsp castor sugar

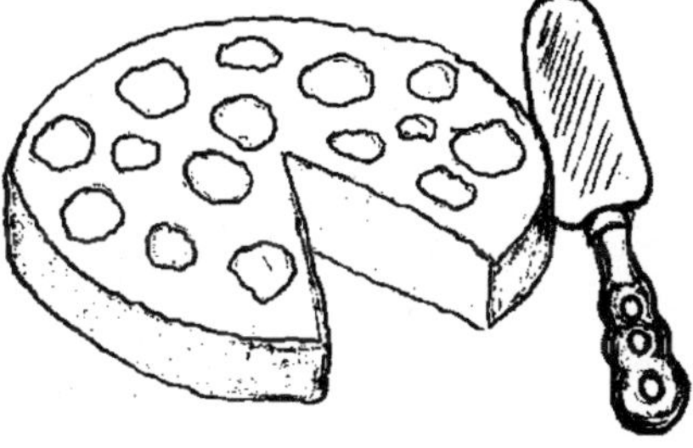

Apricot tea cake

Method

1. Preheat oven to 320°F (convection oven), grease a 8 in cake tin and line base with parchment paper.
2. Whisk egg with sugar until creamy. Add flour, baking powder, salt, milk, zest and juice of the lemon, and melted butter, and blend together until smooth. Pour into cake tin.
3. Arrange apricot halves on top of mixture. Place in oven and bake for 30–40 minutes.
4. Test with skewer to ensure cooked. Leave in tin for further 10 minutes before turning out onto rack. Brush top with melted butter and sprinkle with cinnamon and caster sugar.
   Eat cake within a day or two — it doesn't keep well.

» DID YOU KNOW?

This recipe uses natural butter. Many brands of margarine contain trans fats, which contribute to cardiovascular disease, excessive "bad" cholesterol (LDL) and plaque build-up on artery walls. Better to use butter, which contains a good balance of omega-3 and omega-6 fats, fatty acids, and vitamins and minerals for good health. Note: not all cholesterol is "bad". Low density lipoprotein (LDL) causes fat build-up on artery walls but high density lipoprotein (HDL) reduces this plaque and is important for good cell and body function.

3. Berry cordial and syrup

Suitable for most berry fruit, such as raspberry, blueberry and boysenberry (even frozen). Can be made in less than 30 minutes, and frozen to make icy popsicles.

Ingredients

6 lb berry fruit (blackcurrants, raspberries, boysenberries, redcurrants, mulberries)
10 cups water
8 cups raw sugar (see method point 2)
1 heaped tsp tartaric acid (double for strawberries, blackberries, boysenberries, redcurrants)

Method

1. Place fruit and water in large pot. Bring to boil and simmer for 15 minutes. Strain to remove pulp. (The pulp can be used for sweets.)
2. Add approximately 1 cup of sugar for every 1 cup of resulting juice in a clean pot. Bring to boil and simmer for 2 minutes. If you have less sugar the cordial will perish and not keep well.
3. Add tartaric acid and blend well. Pour into sterilized bottles and seal immediately.

As a cordial, usually use 1 part cordial to 5 parts water. As a syrup, use undiluted over ice cream or other sweets. Refrigerate after opening.

Berry cordial

» DID YOU KNOW?

We should try to avoid concentrated fruit juices, such as orange, apple and pineapple. These tend to have no fiber and lots of sugars, which are easily converted to fats in our body. Excessive sugars contribute to diabetes and other diseases. While we do need to use fair amounts of sugar to make cordial and jam, drink and eat these in moderation.

4. Meat patty

Made from good quality (premium) ground beef. Don't buy the very low fat type as you need some fat to keep the meat moist. Add breadcrumbs and an egg to bind the meat.

You can make hamburgers, meat loaf, spaghetti and meatballs, pasties, and if you add some ground sausage then sausage rolls.

Ingredients and method

The general mix is 1 egg and 1 medium onion for every pound of meat. You can also add 1 tsp celery (or other) salt and 1 crushed garlic clove. Add ½ cup breadcrumbs and hand mix thoroughly. Break into palm-sized (tablespoon) portions. Roll flat to make hamburgers or into balls to make meatballs.

Meat patties

5. Raspberry jam

From start to finish it's about 20 minutes. Raspberries are very high in fiber and vitamin C and good source of vitamin K and manganese, but also high in sugars (calorie-dense). Overall, reasonably nutritious and filling.

Jams tend to have an even consistency, while "conserves" typically have evidence of some chunky fruit pieces. Other jams are discussed in the next section.

Ingredients

3 lb raspberries
¼ cup water
Juice of 1 lemon
3 lb raw sugar

Method

1. Place raspberries, water and lemon juice in a large pot. Heat and gently boil for 10 minutes.
2. Add sugar, stirring continuously and boil for another 10 minutes.
3. Stand for 10 minutes and pour into sterilized jars.

Note: to sterilize your jars, wash thoroughly, dry and place into oven at 225°F for 20 minutes. Turn oven off and allow to cool enough for handling. Place any jams, chutneys and relishes into the warm jars. Screw the lids on. You will find that as the mixture cools it creates a small vacuum to seal properly, although current recommendations suggest processing in a boiling-water canner.

Raspberry jam

» DID YOU KNOW?

Some fruits, such as citrus, quince and cranberries, have high levels of pectin and set easily. Adding lemon juice in jams helps extract the pectin in fruit, whilst also making it a bit tangy.

Raspberries only have medium pectin levels, so if your jam doesn't set too well, you can boil down to make thicker or add a commercial product.

6. Spelt bread

You can use all spelt flour in this recipe if you prefer — it will result in slightly coarser textured bread. The bread is delicious fresh as well as toasted, and spread with butter and honey. While this is a fermented food, it is discussed in this section as one of Sally's top ten.

Ingredients

2 cups organic spelt flour
2 cups organic plain flour
4 tsp dried yeast
3 tsp raw sugar (or 2 tsp honey)
2 tsp salt
2 tbsp light olive oil (or similar vegetable oil)
1–2 cups warm water

Spelt bread

Method

1. In a large bowl, mix the spelt, plain flour, yeast, salt and sugar. Make a well in the center and pour in the oil and almost all the water and mix to a soft dough. (You may only need about ¾ cup of water.)
2. Cover with a tea towel and leave to rise for about 1 hour or until approximately doubled in size. At this stage you can take the dough to the following step, or just turn it over with a spoon and let it rise again.
3. When ready, turn dough out onto a lightly floured surface and knead until smooth. Shape into 2 equal sized balls and place side by side in the tin. Cover with a tea towel and allow to rise almost to the top of the tin.
4. Bake at 400°F for 40–45 minutes until well-risen and golden, and when the loaf sounds hollow when tapped with the knuckles. Turn out onto a wire rack to cool.

» DID YOU KNOW?

Spelt flour is made from the spelt grain *Triticum spelta*, a relative to common wheat *Triticum sativum*. It does have less gluten and more protein than wheat, but is a lower-yielding crop with tough husks around the seed that makes it more difficult to harvest. Some people classify spelt as a subspecies of wheat because in many ways they are so similar.

7. Sweet chili sauce

So easy to make and a great addition of flavor to lots of dishes.

Sweet chili sauce

Ingredients

0.5 lb large red chilies
3½ cups sugar
3 cups vinegar (white or cider)
1 tbsp grated green ginger
10 garlic cloves (crushed and mashed)
1 tsp salt

Method

1. Place all ingredients in saucepan. Bring to boil and simmer for 30 minutes.
2. Allow to stand for 30 minutes and pour into sterilized jars.

8. Tomato relish

This is a more savory dish.

Ingredients

2 lb tomatoes
2 onions
1 red chili
1 red pepper
1 tbsp salt
1½ cups raw sugar
1½ cups cider vinegar
2 tsp mustard powder
1 tsp turmeric
2 tsp tapioca starch

Method

1. Chop tomatoes, chili, onions and pepper into ½ in square pieces, and place in large bowl. Mix salt in and allow to stand for 2 hours.
2. Drain all liquid off. Put mixture into saucepan, add sugar, vinegar, mustard and turmeric and bring to boil. Cook for 30 minutes.
3. Mix the tapioca starch with a little bit of (extra) vinegar to make a paste. Add to the mixture and stir thoroughly. Boil for another 3 minutes (no more or mixture will stick to the pan). Spoon into sterilized jars.

Tomato relish

9. Zucchini pickle slices

Pickles are preserved foods, typically in brine (salt) solution or vinegar. Spices can be added. Normally soak vegetables for a few days in solution, cover with a cloth. Pour off and then refill, and leave for a few days more. Pour off solution and refill one more time and cap.

Ingredients

2 lb zucchini
1 red pepper
2 onions
½ cup salt
2 cups raw sugar
2 tsp mustard seeds (powder will do)
3 cups vinegar (white or cider)
Options: 1 tsp dried chili flakes, curry powder or 1 bay leaf

Method

1. Cut zucchini, pepper and onions into thick slices (½ in), and place these in a large bowl. Add salt and enough water to cover. Mix to dissolve salt. Leave to stand for the day or overnight and then drain.
2. Combine sugar, vinegar, mustard (and optional chili, curry or bay leaf) in a saucepan and continuously stir as you bring to the boil. Add drained vegetables, bring back to the boil and then remove from heat.
3. Spoon into sterilized jars and seal. Store for at least one week to enable vinegar and flavors to penetrate vegetables before you taste. Will store well if kept in a cool, dark place.

Pickled zucchini slices

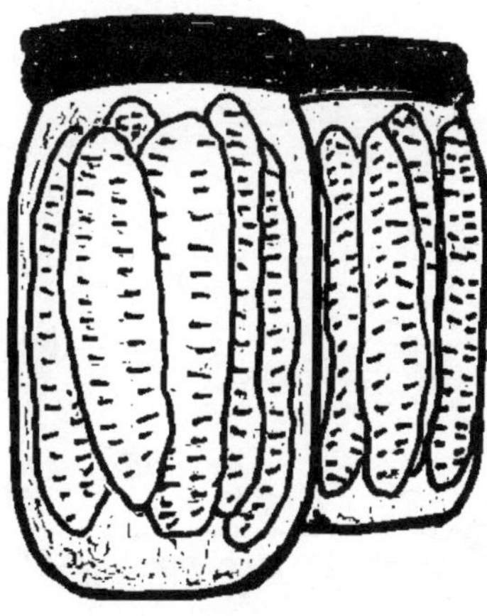

10. Zucchini slice

Zucchini is underrated and does contain high level of vitamin C as well as a fair range of vitamins and minerals (up to 10% of daily allowances). While zucchini has very little fat and cholesterol, it still does have reasonably high carbohydrate content from sugars.

Ingredients

5 eggs
14 oz grated zucchini
1 onion, chopped
6 oz bacon
½ cup chopped semi-dried tomatoes (option: grated carrot or pumpkin)
1 cup grated cheese
⅓ cup diced basil leaves
1 tbsp chopped parsley
¾ cup flour
1 tsp baking powder
½ tsp salt
½ cup light olive oil

Method

1. Preheat oven to 350°F. Grease and line a baking tray.
2. Whisk eggs in a large bowl and add all other ingredients. Pour into tray and bake for 30 minutes (or until slice is firm).
3. Allow to cool for 5 minutes before cutting into squares.

Zucchini slice

Others to try

I will admit it. These are my favorites, but, in my defense, they are both easy to make (because I can!) and are nutritious.

1. Fig jam

While this specifies figs, almost all other fruits utilize the same recipe. You could experiment with less sugar and add lime juice, or add two different fruits together.

Fig jam

Ingredients

2 lb figs
2 lb sugar
Water
Juice of 1 lemon

Method

1. Chop figs into slices. Place in pot and just cover with water. Add sugar, heat and stir to dissolve.
2. You will need to stir every now and again to prevent burning of the mixture on the bottom of the pan.

2. Garlic salt (or flavored salt)

A simple change to make ordinary salt extraordinary. You can add 1 tsp of garlic powder (or any other herb powders) to every 10 tsp salt to flavor it.

Garlic salt

3. Lemon curd

Great sweet filler for tarts, lemon meringue pies, with ice cream, spread on toast and the list goes on. This recipe makes about 1½ cups of curd, which is also known as lemon butter.

Ingredients

2 eggs
½ cup sugar
¼ cup lemon juice (one large lemon)
1 tbsp grated lemon zest (from same lemon)
½ cup butter in finely chopped pieces

Method

1. Place eggs, sugar, lemon juice and zest in a saucepan.
2. Heat gently and whisk continuously.
3. Whisk until curd becomes thick. Do not boil curd. Remove from heat and leave to cool. It should thicken further.

» DID YOU KNOW?

Sugar is used in almost every jam recipe. It is not so much that sugar makes the jam sweeter (it does) but it acts as a preservative and prevents microorganisms from growing (they can't survive in high sugar as their bodies lose water and die).

Sugar, along with a little acid, also provides the right medium to enable pectin (a polysaccharide like starch and cellulose) to change into a gel form. If you make a jam with little or no sugar, it will be runny and you need to keep it in the refrigerator and use it within a few weeks.

4. Powdered lemon

A good-producing lemon tree will yield far too many lemons for a family to use or give away. Some lemons can be stored in a couple of practical ways.

Firstly, you can freeze them. But this does damage the tissue and when they are thawed out they don't look that great. They often expand and split during the freezing process, enabling juice to be lost.

Secondly, you can cut the lemons in slices and then dry in a dehydrator. Once dried, the slices can be stored in dark jars or paper bags until needed. They seem to keep the flavor well and when added to dishes the lemon flavor is evident.

Finally, the next step is to put the dried slices into a blender and grind into powder. Make sure you remove the seeds first! The lemon powder keeps well, as long as it is kept dry.

5. Quince jelly

Ingredients

2 lb quince (4–5 large)
2 lb sugar
Juice of 1 lemon
6 cups water

Method

1. Peel quinces or rub to remove hairy skin off fruit. Slice and place in a saucepan. Add lemon juice, water and about half the sugar.
2. Bring to boil and simmer for up to 1 hour. Strain through colander but collect juice. The pulp should have pink tinge, and is eaten as a sweet. The liquid can then be used to make jelly.
3. Put the liquid in a pot back on stove, add remaining sugar and stir while bringing to boil.

   Gently boil down until setting point is reached. This is determined by placing a drop or two on a cold plate and seeing if it sets like jelly. When this occurs it is ready.
4. Pour into a shallow plastic tray or ice cube tray, so you can easily cut or remove pieces when required.

» DID YOU KNOW?

Quince trees are unusual. They belong to the apple and pear family of plants, but the fruit are too woody to eat raw. They do need cooking to soften the flesh and make palatable.

This section is not a substitute for a good cookbook, and while I have tried to provide recipes for healthy food, there are ingredients that some would argue aren't that healthy.

Generally, high amounts of sugar (carbohydrates) are not good for our health, and most people know that low-carb fruit and vegetables are much better for human health.

Furthermore, overcooking destroys lots of the goodness and nutrition, so raw foods are better for this. A lot of the ingredients are gluten-free but many are not. You can easily source gluten-free flour and other gluten-free ingredients if you need them.

I remember reading somewhere that "you can't exercise your way out of poor nutrition," so clearly eating better quality food is important, but it may not solve your health problems.

It is all about balance and moderation, and that applies to all aspects of life.

Ten fermented foods

Fermented foods are becoming more popular as more and more people grow some of their food and people want to make more connections (and reconnections) to their food. The next obvious step is to preserve some of their excess produce, and fermenting foods is just one way this can happen. However, many people are cautious at the thought of eating or drinking fermented foods. Concerns about botulism, poisoning, fungi and molds, and pathogenic bacteria are largely unwarranted. If you make a fermented product and it doesn't look right or smell right, you instinctively know to throw it away.

Almost everyone eats fermented foods every day or so as they include cheese, beer (and ciders, sparkling fruit drinks, ginger beer), black tea, soy sauce, yogurt and vinegar.

Then there are the more obscure or less common products, such as sauerkraut (and many other fermented vegetables), sourdough, miso, kefir grains, tempeh and kombucha.

Fermentation is basically anaerobic digestion by primarily bacteria and fungi (including yeasts), so excluding air is essential for the process to occur. When microorganisms feed on the sugars in the foods they digest, alcohol, acetic acid (vinegar) and lactic acid (think sour milk) are produced and all of these substances inhibit the growth and action of other (harmful) microorganisms. The fermented food becomes preserved food.

However, fermented food doesn't last forever, and most products are usually stable for a week or two, and occasionally longer, depending on the pH, salinity and temperature of the fermented food. You need to look, smell and taste as you go along this path, and keep in mind that fermented foods should be seen as just one part of our dietary mix.

What you need to start and what you need to be aware of

1. Temperature

Some cultures can work at room temperature, but many require temperatures around or slightly higher than that of our own body temperature (so many cultures thrive between 98 and 113°F), so it is crucial that somehow you maintain a reasonably constant temperature during the fermentation process. You can do this by using a water bath, large thermostat-controlled pot or urn, a haybox cooker, thermos flask or something to keep your culture warm.

» DID YOU KNOW?

Probiotics is a term that describes the useful live microorganisms that we ingest to provide us with some health benefit.

Our gut contains millions of living microorganisms already — many essential to help us digest foods so that these can be assimilated into our bodies — but sometimes the microbial activity in our digestive system is out of balance. By eating what is essentially a live culture, microbial feed supplement, some benefits to particular health problems such as diarrhea, gingivitis and gastritis seem to occur.

The scientific research into specific strains of microorganisms and their effect on our health is ongoing.

You also need to obtain a milk or beverage thermometer (not mercury or alcohol types), preferably digital but analog types tend to be cheaper.

2. Jars and bottles

Obtain a selection of bottles and jars.

A selection of different sizes, with seals, will make your job a lot easier. Mason and Kilner jars (wide mouth, with screw lids), storage jars (clip top with rubber seal) and swing-top bottles (with rubber Grolsch seal) are all required to preserve your products.

3. Cultures

Every fermented product requires a specific microorganism culture. You have a choice with cultures: either obtain a known (mostly bought) culture or let nature take its course (a "wild ferment") where a variety of bacteria and yeasts in the air inoculate the mix.

4. Sundry equipment

Cheesecloth, muslin, balance or scales. Bucket chemistry is alright some of the time, but proper ratios of ingredients ensure better results. Muslin and cheesecloth are both fine cotton cloths (but do come in various mesh sizes), but any fine woven cotton material will work — even a tea towel.

5. Hygiene

The three rules for success of any fermented food are: 1. cleanliness, 2. cleanliness and 3. cleanliness. I think you get the point.

Here are some common, well-tried, easy-to-make fermented foods, listed alphabetically.

1. Cheese

Cheese is usually made from milk and you can use any animal milk, such as that from cows, goats and sheep. Soft cheeses are easy to make; hard cheeses are a little more complicated.

Soft cheese from milk

Ingredients

4 cups milk
2 tbsp vinegar (or lemon juice or 1 tbsp rennet will do)
Option: pinch of salt

Method

1. Heat milk to about 100°F (or 175°F if you are using raw or unpasteurized milk) in a saucepan, constantly stirring with a wooden spoon so the milk doesn't scald.
2. Add the vinegar and mix thoroughly. This will turn the milk into curds (solid part) and whey (liquid part).
3. Allow to cool to room temperature and then pour the mixture through a muslin cloth. Squeeze the cloth to remove all whey. I sometimes add another tbsp of vinegar to the whey because you usually get another small amount of curds that you can filter out.
4. Scrape all curds into a bowl. Add salt and gently blend in.
5. Place in a refrigerator. Eat within 3 days, after which it tastes tangy and is better suited for recipes that require cheese in cooking.
6. To make into a block of cheese, wrap the cheese in a piece of muslin or cheesecloth, place in a shallow dish and then add a weight on top to flatten (another dish, a block of wood). You need a shallow dish to contain the small amounts of whey that continue to be squeezed out. Leave in refrigerator for a day, unwrap and cut into cubes.
7. Don't expect a large amount of cheese from 4 cups of milk. About 2.5–3 gal of milk are required to make about 2 lb of cheese, so 4 cups makes less than 4 oz.

Hanging cheese in a refrigerator

Soft cheese from yogurt

Ingredients

2 lb plain yogurt (must contain live cultures)
½ tsp salt
Flavoring: ½ tsp of either finely-chopped rosemary or basil leaves or lemon zest or ground black pepper

Method

1. Mix all ingredients together in a bowl.
2. Spoon mix into a double layer of muslin or cheesecloth. Tie into a ball with string and suspend in a large bowl in the refrigerator or a very cool place for 2 days. The whey should drip out into the bowl and can be used elsewhere or discarded.
3. Remove the solids and roll to small balls, so these can be eaten as required, although in many cases the cheese is similar in consistency to cream cheese.
4. If kept in a refrigerator, the cheese balls should last a couple of weeks.

» DID YOU KNOW?

Milk is made up of a protein part, which separates into "curds," and a watery part, which is the "whey." The whey contains water-soluble minerals, lactose, proteins and many beneficial microorganisms that can be used to "seed" other cultures when fermenting vegetables and beverages.

Most of the proteins in milk are known as casein proteins. Curds form when lactic acid causes the casein proteins to coagulate (curdle) together.

Simple hard cheese

The difference between making a soft cheese compared to a hard cheese is that you add a proper culture to make the flavor you want and you need a cheese press to really compress the curds. Because hard cheeses are left to "mature" for some time, it is also important to ensure all containers and implements are sterilized — usually by boiling in water for 5 to 10 minutes.

Ingredients

4 cups milk
½ tsp salt
Cheese culture (mesophilic — room temperature variety)
Rennet or vinegar

Method

1. Gently heat milk in a pot (or water bath) to 85°F.
2. Add the bacteria culture, stir in and try to maintain the warm temperature for about 30 minutes. Monitor your thermometer as overheating will kill the culture.
3. Turn off the heat and allow the curds to coagulate (clump together). Cut the curds into small chunks.
4. Now reheat the curds and maintain a temperature of about 75–100°F for another 30 minutes. Reheating the curds forces more whey out.
5. Allow the mixture to cool to room temperature. Strain the curds through muslin, sprinkle the salt on and gently blend together.
6. Scrape the curds into a container that has a small hole or two on the bottom (this allows whey to drain away). Place the container into your press, add weights and leave for an hour or so.
7. Remove weights, open up press, gently flip cheese over and re-install press and weights. Keep pressed overnight.
8. Next day, unwrap cheese and keep in a cupboard or pantry to keep dust off and so on. You could turn every second day. As it matures the taste changes. Different types of cheese require different maturation times. After a few weeks try a piece to see if that's the taste you like. Some cheese surfaces are sealed by melted wax, so molds don't grow on them.

» DID YOU KNOW?

Rennet is a mixture of enzymes produced in the stomachs of ruminant animals such as cows, sheep and goats. These enzymes act on the proteins (casein) in milk and cause it to curdle (to become the curds).

Normally you need to use the specific rennet from an animal to work on their milk, so calf rennet works best on cow's milk.

There are also vegetable rennets, and these are derived from particular molds and extracts from plants such as figs, capers and thistles. Acids such as citric acid (lemon juice) and acetic acid (vinegar) will also curdle milk if you cannot obtain a suitable rennet.

2. Cider

Apple juice can be fermented and made into cider, a slightly alcoholic drink with a sour taste. As always cleanliness and sterilizing bottles, equipment and buckets is essential.

Ingredients

20 lb apples
1 package wine or ale yeast
Sugar

Method

1. Compress diced apples by squashing or blending to extract juice. Use a press if you have or can make one. Add the juice into a large bowl or bucket. You will need a lot of apples to make a few cups of juice. 1 gallon is enough.
2. Add yeast powder and mix thoroughly. (You may like to try with a wild ferment — just letting the natural yeasts in the air start the process, but the results tend to be unpredictable.) You don't need to add any additional sugar at this stage as most fruits have high sugar content already. Sugar is digested by the yeasts and makes carbon dioxide gas for the fizz. It also makes alcohol, so the more sugar you put in the higher the alcohol content of your cider.
3. Cover with cheesecloth or a cotton cloth and leave to ferment for a week or so. Taste the cider. The longer it ferments the more sour it gets so

stop the process when you like the taste. Ideally, yeasts work in anaerobic conditions so you could transfer the brew into a flagon or demijohn and insert a fermentation lock (water trap) to allow carbon dioxide to escape but prevent oxygen in the air from entering. Usually after a couple of weeks the yeasts die off and the cider clears.

4. You can now bottle the cider. Decant the clear fluid and try to leave the sediment behind. Even so, filter the juice through muslin and pour into bottles that can be capped. Use bottles made for sparkling wines or PET bottles would suffice.
5. To each bottle add 1tsp sugar. This will re-activate enough yeast to make carbon dioxide to make the drink fizz when you open it.
6. Keep the bottles in a warm place for at least 3 days so that the yeast can ferment the sugar. After this time you can place in the refrigerator or a cool place for storage.
7. Be mindful that when you open the bottle, gas will rush out and there will inevitably be some sediment on the bottom.

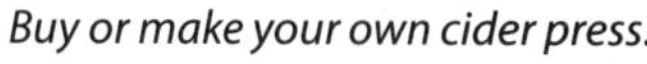

Buy or make your own cider press.

A fermentation lock prevents air from entering.

3. Fermented vegetables

This is the basis of dishes such as sauerkraut and kimchee (both based on cabbage) but any combination of vegetables will work. Cabbage leaves and root crops seem to work well.

Ingredients

- 1 cabbage (or other leafy brassica)
- 1 beetroot or large carrot (Koreans add radish and hot peppers for kimchee)
- 1 tbsp salt (general rule of thumb is 2 tsp salt for every 1 lb of vegetables)
- Spices to suit — to add flavor, e.g. lemon zest, sliced ginger or garlic, and fennel or dill seeds
- Option: 1 tbsp whey

Method

1. Dice all vegetables. Sprinkle salt over them. At this point you could leave in the bowl for a couple of hours as water is drawn out of the vegetables, and this sometimes makes packing into jars easier.
2. Push them tightly into a glass jar, with the help of a wooden spoon. The salty brine solution, made from the juice of the crushed vegetables, should cover the vegetable mix. You might need to add a little extra water to ensure the vegetables are completely covered. Adding whey (optional) often kick-starts the ferment.
3. Place a lid and weight inside the jar to keep the vegetables below the juice.
4. Cover the jar with a cloth to prevent insects from entering.
5. Leave the ferment for a day or two and then check. It should start to taste tangy and sour. Leave for a few more days if you want it to become more sour.
6. Store in refrigerator. Use over a few weeks.

Sauerkraut in a clip-top storage jar

4. Ginger beer

This is easy to make, but you need to make a "ginger plant" first, and culture this over a week.

Ingredients

Ginger — to make 5–7 tsp full of grated ginger
Sugar — similar volume
Option: ½ tsp of dried yeast
(More sugar and water later)

Method

1. Place 1 cup of warm water (100°F maximum) in a glass jar.
2. Add 1 tsp of grated ginger and 1 tsp of sugar and stir to dissolve sugar. You can add the dried yeast at this point but the ferment will work with wild yeasts from the air.
3. Cover with a cheesecloth and place jar on a windowsill, so that it stays warm.
4. Every day for a week (5 days minimum), add 1 tsp grated ginger and 1 tsp sugar.
5. After a few days you should notice foaming and bubbles. After one week strain the "ginger plant" through several layers of cheesecloth over a bowl and collect the resulting liquid.
6. To make the ginger beer drink add 2 lb of sugar to 1.5 gal of warm water in a clean bucket. Stir to dissolve. Mix in your ginger plant juice.
7. Pour the mixture into bottles and cork. You can use glass bottles but the gas produced from fermentation might cause the bottles to explode. Screw-top plastic (PET, common soft drink) bottles may be a better option, as when you see them swelling a little you can unscrew the lid to relieve the pressure and then tighten again.
8. Store bottles in a cool place and allow further fermentation for 3-5 days. Place bottles in a refrigerator to stop fermentation process.
9. Use the sieved "ginger plant" to make a new culture. Place ¼ to ½ of the residue "plant" you captured in the cheesecloth and add this to 1 cup of warm water. Add fresh grated ginger and 1 tsp of sugar and repeat this whole process to make more ginger beer. Give the other fractions of the ginger residue to friends so that they can start their "plant" too.

Ginger beer starter

5. Milk kefir

Kefir grains are not "grains" in the true sense but are clumps of particular bacteria and yeasts that live symbiotically together. There are basically two common types — milk kefir and water kefir, and they are not interchangeable.

Ingredients

2 tbsp milk kefir grains
4 cups milk

Method

1. Add kefir grains to milk in an open-mouthed jar.
2. Cover jar with a cloth and leave on the bench for a day (after 6 hours it should start to taste sour, so it can be eaten any time after this).
3. Pour mixture through strainer to remove the grains.
4. Keep the grains in a refrigerator to use at another time, and the resultant fermented milk can be added to fruit and blended to make smoothies.

Milk kefir grains

6. Sourdough bread

Sourdough bread is fermented by wild yeasts in the air. You initially culture a starter, much like ginger beer, and then use this to make the loaf of bread.

Ingredients for starter

9 oz plain flour — can be organic, white, wholemeal, gluten-free, rye, a combination of these — your preference, but fully wholemeal or rye loaves are quite heavy and don't rise as much, so best to mix with white flour in the ratio 1:1.

9 oz water

Method for starter

1. Initially mix 3 oz flour and 3 oz of water in a bowl to make a sticky paste. Cover with a damp tea towel.
2. Leave on the kitchen bench for 2 days. (Check the tea towel — keep it moist.) The dough should look bubbly — if no evidence of bubbles then leave for another day or so before feeding.
3. Each day for the next 2 days (days 3 and 4), mix in 3 oz of flour and about the same of water to make a soft dough. It shouldn't be too runny, so maybe less water sometimes.
4. Divide the sourdough into two. You only need about 6 oz dough to make the bread; the other half is used as a starter for another loaf.

Place this new starter in the refrigerator and feed every few days with a little flour and water to keep alive if you are not going to make another loaf straight away.

Ingredients for bread

2 cups flour

1 tsp salt

1 cup water, plus extra as required

Method for bread

1. Combine flour and salt in a large bowl. In a separate bowl add 1 cup water to the sourdough to make it runny.

2. Pour the sourdough mixture over the flour and mix by hand. Add water, or a little flour, as required to make soft dough — not crumbly nor sticky.
3. Knead the dough on a lightly floured bench or plastic kitchen mat for about 10 minutes. Flour your fingers and hands too and work the dough until you find that it is elastic.
4. Return the dough to the large bowl and cover once again with a damp tea towel. Let it rise overnight (it should double in size). Popping bubbles means it is more than ready for the next stage.
5. Use your knuckles to knead the dough once again. It will decrease in size. Place back in bowl, cover with damp tea towel and let rise again over 5 or 6 hours.
6. Turn the dough onto a baking tray. Gently shape it a little to make a cylinder and slash the top once or twice with a knife to vent any water as it cooks. Cook in preheated oven at 425°F for about 30 minutes. Check to see if it sounds hollow when tapped and is firm. Remove from oven and turn onto a wire rack to allow the bread to cool.

» **DID YOU KNOW?**

The addition of salt is an important step in many fermentation processes. The salt slows the rate of fermentation by yeasts and bacteria.

Do not add it at the start of some recipes as you won't get the fermentation working. So for sourdough, you add the salt to the dough to be baked but never to the starter.

7. Sparkling cordial

It's so easy to make and much better than the soft (cool) drinks you buy as there are no additives. Elderflower makes lovely cordial, and so do rose petals (it's like pink champagne).

Can also use rhubarb, strawberry, raspberry, cherry and plums (apples make cider).

Ingredients

2 cups rhubarb (or other fruits, or 6 elderberry flower heads, 2 cups rose petals)

2 cups sugar

Juice of 1 lemon

⅓ cup white wine vinegar (or cider vinegar)

15 cups water

Method

1. Combine all ingredients in a clean bucket and cover with a tea towel. (You might need to dissolve sugar in a small amount of hot water or stir well to dissolve.)
2. Leave to ferment for 48 hours.
3. Strain and pour into plastic bottles and seal. Wait 1–2 weeks before opening.
4. Crack the cap to see if fizzy and ready for drinking. Keep outside just in case of explosions.

Sparkling cordial

8. Vinegar

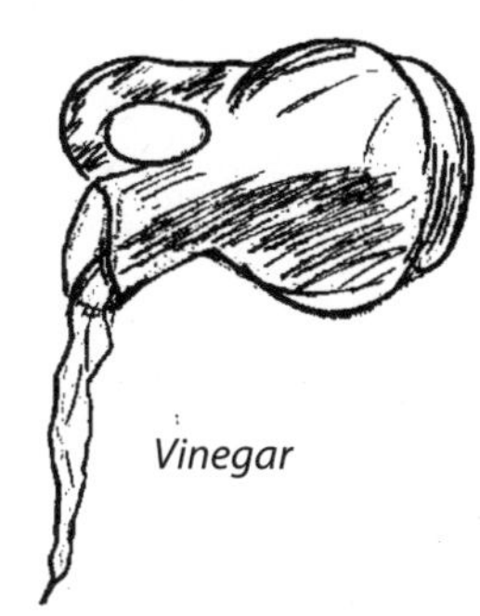
Vinegar

Vinegar is made from fermented alcoholic drinks such as apple cider and some of the sparkling cordials. Bacteria change ethanol into ethanoic (acetic) acid in the presence of air, so this is aerobic digestion.

Like sourdough, you need a starter culture, which is called mother of vinegar — a harmless slimy mass of bacteria and cellulose that you can buy or is made during the vinegar process.

Ingredients

Alcoholic drink such as cider, wine, fermented fruit juice. Make sure your wine or juice does not contain preservatives, as the vinegar won't begin to form. Check the label.

Method

1. Place the fermented liquid into a bottle or jar. Cover with a cheesecloth to keep dirt and insects out.
2. Keep bottle in a dark place. Fermentation will occur faster in a warm place (75–85°F). A gelatinous layer (the mother of vinegar) forms in the surface. Leave for about 1 month.
3. Test the liquid to see if vinegar is forming. Smell and taste. It might be ready at this stage but often it takes a few months for complete conversion.
4. When you think it is ready, filter to remove the mother of vinegar (which can now be used to make more vinegar, quicker than before). Bottle and cap vinegar, and keep in cupboard or refrigerator.
5. Option. You can pasteurize vinegar if you are concerned about excess bacteria. Heat the vinegar to about 175°F for 10 minutes as this will both drive off any excess alcohol and kill all microorganisms. However, be warned, heating will diminish the flavor of the vinegar.

» **DID YOU KNOW?**

Balsamic vinegar is a specialized vinegar made from grapes, then aged in wooden barrels for many years.

Caramel or other flavorings may be added, and after heating to concentrate the ingredients and further aging, a dark sweet and sour vinegar is produced. Not usually made by homeowners due to the length of time the whole process takes.

9. Water kefir

Like milk kefir, water kefir grains are a special blend of bacteria and yeasts that thrive in nutrient-rich, watery environments.

Ingredients

¼ cup water kefir grains
¼ cup sugar
Juice of 1 small lemon
2 roughly-diced (chunky) figs (dried or fresh)

Method

1. Heat 8 cups of water in large pot and bring to the boil.
2. Stir in the sugar to dissolve.
3. Allow to cool and pour solution into 2 or 3 large jars. Add ⅛ cup of kefir for each 4 cups in a jar — you can work out the approximate proportion. Add the equivalent pieces of 1 fig and 1 tsp lemon juice for each 4 cups too.
4. Cover with cheesecloth and allow to ferment for a few days. Taste — the longer it ferments the stronger the taste will be.
5. Strain mixture, discarding the spent figs, but keeping the water grains for another time, and collect the juice. Re-culture the grains and store in a refrigerator.
6. The juice can be drunk anytime, kept in the refrigerator until needed, or you can ferment it even further.
7. For a secondary ferment, pour ¼ cup of any fruit juice or 1 tbsp sugar into a 28 oz swing-top bottle.

   Fill with water kefir until very near the top. This will enable you to produce a flavored kefir drink or one that is fizzy.
8. Allow the bottles to ferment for another day or two on the kitchen bench. Store in refrigerator until needed. Be careful when opening as gas pressure may cause contents to shoot out.

Water kefir

10. Yogurt

Fermented dairy products can be easier for us to digest because the sugar (lactose) has already been broken down and the complex protein casein has also been broken down into amino acids.

Ingredients

4 cups milk

1 tbsp plain yogurt culture (must be live culture)

Method

1. Heat milk to about 104–113°F (if you think milk may be contaminated, then heat to 176–185°F to pasteurize it, but let it cool to 104°F again).
2. Add yogurt culture and stir in.
3. Keep mix warm by placing in a thermos flask or wrapped in something to insulate the mix.
4. Test the mix after 6–9 hours. The longer you leave it, the more sour it gets. Once you are happy with the flavor, keep the yogurt in the refrigerator.

Heating milk is important to denature the proteins — this treatment stops the yogurt staying lumpy.

If you want to thicken the yogurt, strain the mix through a cheesecloth and either discard or keep the "whey" that passes through. To flavor yogurt, add chopped fruit, honey or vanilla essence.

Don't be discouraged if you don't have immediate success with these fermented food projects.

It took me a few times to get the sourdough so it didn't look and taste like a rock, but the lemon curd, yogurt, soft cheese and a host of others worked perfectly the first time, and still continue to do so. Persevere.

Yogurt

Preserving your foods

For the home gardener, foods can be most often preserved by pickling, salting, smoking, drying (dehydrating), bottling, freezing (and refrigerating) and preserving in alcohol, syrup or some other solution or medium.

So far in our discussion about foods you can make at home we have discussed drying (powdered lemon), preserving in a sugary medium (raspberry and fig jam), preserving in alcohol (cider), pickling (chutney and relish in vinegar) and bottling (ginger beer, sparkling cordial).

Drying food

Vacola canning jars

We have also talked about making garlic salt (for salting food) and vinegar (for pickling). Not to mention refrigeration for most of the fermented foods (which themselves are partially preserved foods).

More commercial, specialized preserving techniques include vacuum packaging, canning, adding chemical preservatives such as sulfur dioxide, and freeze drying. These are not discussed here.

Smoking of red meat and fish both preserves and flavors the food, but you generally need to build a smokehouse, have a good fuel supply and time to do this properly.

However, excess fruit and vegetables can easily be dried or canned. Canning invariably uses heat to kill microorganisms, to drive off air by the action of boiling water, and then sealing in a partial vacuum as the solution cools down.

Drying drives off almost all moisture so that microorganisms cannot survive or breed.

IT'S ENERGY THAT MATTERS 15

Energy is essential to all life. We rely on the Sun to drive all ecosystems on Earth and we eat food simply to obtain energy for us to function. Light and heat energy are commonly known as these are provided by the Sun, but there are many other forms of energy that include chemical, potential, mechanical, sound, gravitational, electric and kinetic.

Some of these energy types are directly and indirectly discussed here as we look at how we can use natural energies to live better and smarter.

Embodied energy

Whenever we need to make decisions about energy we should undertake an energy analysis. We need to consider the embodied energy of materials and their life cycle assessment.

Embodied energy is the energy used in the production of a building, a product or some other structure. The amount of energy used in the manufacture of these materials can be a significant component of the life cycle impact of a home, for example. While it is important to improve the energy efficiency of any building, the amount of embodied energy it contains can be equivalent to many years of its operation.

The importance of embodied energy, and the environmental impacts of using some particular building materials, only becomes apparent when we examine the materials from a life cycle approach, usually known as Life Cycle Assessment (LCA).

LCA examines the total environmental impact of a material through every step of its life — from obtaining raw materials to using it in your home and then disposal or recycling. LCA can consider impacts such as resource depletion, energy and water use, greenhouse emissions and waste generation.

The amount of embodied energy varies with different construction types. A higher embodied energy level can be justified if it contributes to a lower operating energy. For example, large amounts of thermal mass, high in embodied energy, can significantly reduce heating and cooling needs in well-designed and insulated passive solar houses.

You could further reduce the embodied energy of the thermal mass if you used concrete made from recycled building material, such as crushed concrete, rock or bricks, for the aggregate.

Generally, the more highly processed a material is, the higher its embodied energy. For example, plastics have much higher embodied energy than sawn timber, aluminum window frames much more than wooden window frames, and clay bricks much more than stabilized earth. In fact, for every square foot of wall, a cement-stabilized rammed earth home has less than half the embodied energy of a double clay brick home.

There are also other considerations when choosing particular building products. For example, high monetary value, high embodied energy materials, such as stainless steel and aluminum, will almost certainly be recycled many times, reducing their life cycle impact. Furthermore, comparing the energy content per square foot of construction is easier for designers than looking at the energy content of all the individual materials used.

Saving and reducing energy

It doesn't take too much effort to save energy or to reduce the energy we consume in our daily lives. We can also adopt simple technologies that rely more on human power than on fossil fuel power.

» DID YOU KNOW?

Recycling 20 aluminum cans uses the same amount of energy required to make one new can from raw materials. Recycling 1 t of steel saves 1.1 t of iron ore, 1,320 lb of coal and 120 lb of limestone. Recycling newspapers for one year can save enough electricity to power a four-bedroom house for four days. The reuse of building materials commonly saves about 95% of the embodied energy that would otherwise be wasted.

Human behavior and actions

We are all aware that the energy we currently obtain from fossil fuels is finite — it will run out one day. Renewable energy is our best option for electricity production in our future, but it has limited application to make a car work all day, every day.

As petroleum becomes scarce there will be major issues with every aspect of transport. As an individual there are some things you can do now to reduce the overall consumption of fossil fuels. Here is a list of 20, but I am sure you could think of lots more:

- Carpool. Join others on shared journeys to the office or shops.
- Take public transport. Reduce your car use by taking public trains, buses, ferries and trams.
- Bicycle. For short distances use a bicycle, even an electric one.
- Buy an electric car, or convert your current vehicle to run on solar power and batteries.
- Choose energy-efficient appliances and white goods, including low wattage light bulbs. Look for the energy rating — choose those that have the most stars.
- Install water-efficient devices and fixtures (such as taps, showerheads, washing machines, toilets) — this reduces the amount of energy government and private agencies use to treat and pump water to your house.
- Double-glaze windows. This can be as simple as putting a film over the glass window panes.
- Switch off appliances on standby — turn off completely when not in use.
- Set the thermostat for any heating and cooling devices to at least two degrees warmer in summer and two degrees cooler in winter.
- Follow the principle of passive solar design for your house as outlined earlier, such as: draft-proof doors and windows, shade windows that are receiving direct sunlight in summer, install insulation in the roof space and incorporate garden structures to help with heat control.
- Adjust your refrigerator setting to be a little warmer, and the hot water system to be a little cooler.
- Wash only full loads in the washing machine. Use a cold water setting rather than hot water for the wash.
- Turn off lights in rooms that you are not using. It is a myth that you use more power turning lights on and off rather than just leaving them on all the time. Flick the switch as you leave the room.

- Install solar tubes or skylights in rooms that are naturally dark, even in the daytime.
- See if you can do without that second fridge on the veranda or at least turn it off until you really have to use it.
- Close all external doors and windows when the heater or air conditioner is running.
- When you have to replace your water heater, install a solar hot water system with a gas booster, a conventional heat pump or better still, a ground-source heat pump or a heat pump that uses solar water collectors to preheat the water.
- Insulate all exposed hot water pipes with lagging or some type of insulation to reduce heat loss.
- Turn a ceiling or stand-alone fan on rather than switching on the air conditioner.
- If wood is readily available use it in a stove to cook food or in a room heater.

» DID YOU KNOW?

When compared to top-loading washing machines, front-loading machines use 50% less water, 40% less energy and up to 50% less detergent.

Energy-efficient housing

Energy-efficient or passive solar houses save both money and the environment. Passive solar means that the Sun's energy is used to heat and cool a home without the use of pumps, fans, air conditioners, wood stoves and other devices to keep us warm or cool. A carefully planned house can be built that will be 10–20°F warmer in winter and at least 15°F cooler in summer.

In a climate-sensible or passive solar design of a house there are six key principles. The optimum living conditions and comfort will be found in houses that contain some combination of these principles when they are designed and built. These principles can be remembered by the acronym TO VIEW — thermal mass, orientation, ventilation, insulation, external influences and window placement. Most of these principles are well documented and generally well known, but here is a brief summary of each.

T — Thermal mass

Certain materials have the ability to store heat. Heavier, dense materials such as concrete, stone, slate and bricks have a high heat capacity per unit volume to store heat. Houses made from brick, rammed earth or concrete have generally more stable temperatures than houses made of timber, plaster or metal, which tend to heat up and cool down quickly.

Dark surfaces are also better at absorbing and radiating heat than lighter materials. Dark slate and tiled floors can absorb winter sunlight during the day and then slowly radiate or give off this heat into the room at night to maintain a constant level of comfort.

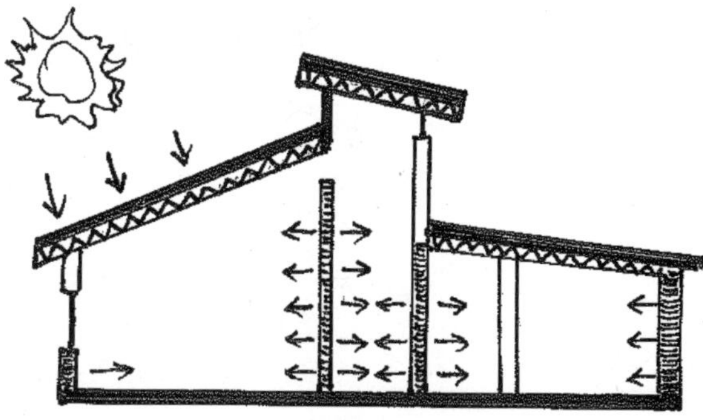

Brick and solid walls can absorb heat during the day and then radiate the heat to keep a house warm at night.

» DID YOU KNOW?

Black objects are the best absorber and radiator of heat. White-colored surfaces are the best reflectors of sunlight and heat. While dark-colored surfaces do absorb heat faster than lighter surfaces, they also give off this heat quicker, and thus cool down faster.

O — Orientation

The correct orientation of your house takes advantage of the Sun's movement during the day and throughout the different seasons of the year, and a good design would consider the Sun's lower position in the sky in winter and a much higher position in summer.

Houses that are rectangular and about twice as long as wide with the long axis stretching east-west tend to remain warmer in winter and cooler in summer. This is because more surface area (walls) face the winter Sun and so can absorb heat. During the summer there is minimum

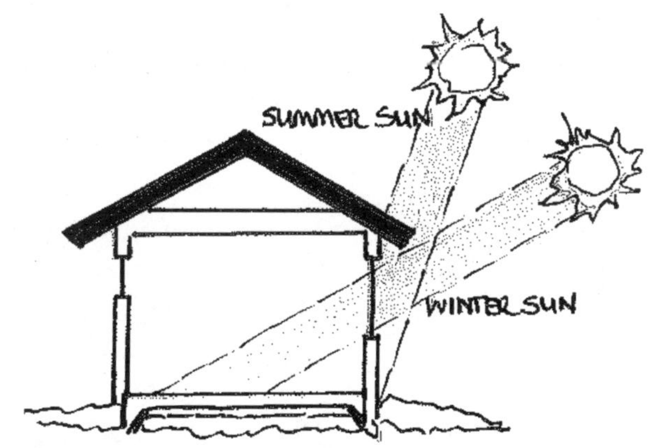

Eaves can shade summer sun but allow winter sunlight to enter a room.

wall exposure to the Sun in the morning and afternoon and this reduces heat absorption, keeping the house a little cooler.

Eaves are an important design feature for houses built in warmer climates. Eave size or overhang can be as little as 20–30 in on the sun-facing side to sufficiently shade walls in summer. More protection is warranted on the east and west sides, so eaves can be larger or verandahs and trellises can be used to shade these sides.

Eaves are not so important for houses in colder climates where there could be too much shading or snow accumulation on the roof.

V — Ventilation

The correct placing of windows ensures good cross-ventilation of cooling breezes. Harvesting and directing breezes in summer can effectively cool the house down without the additional use of air conditioners.

During winter, doors in unused rooms should be closed and any gaps under the door sealed, as leaky buildings lose considerable heat.

Ventilation of the roof space should also be considered. Roof ventilation will allow insulation to work more effectively by allowing hot air a means to escape. However, vents should be closed in winter and cold weather to retain heat.

I — Insulation

Ideally, walls, floors and ceilings should be insulated.

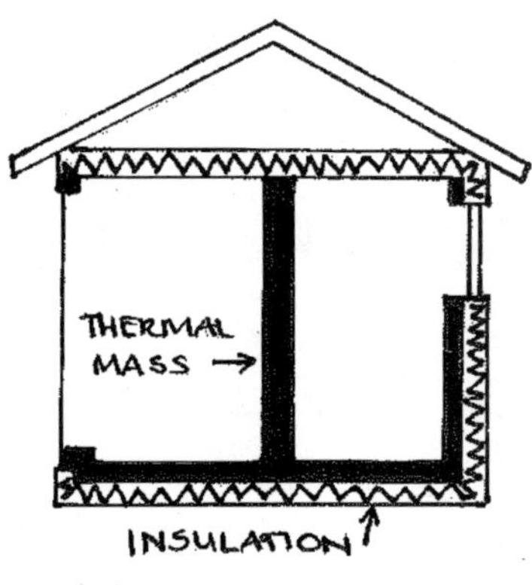

A range of materials can be used to insulate your home and prevent heat transfer. Heat is easily lost and gained through walls and ceilings, so at least these should be insulated. Solid walls transfer heat and cold, so building cavity-walled houses enables either air or some other material to be placed in the cavity to reduce thermal conductivity.

Common insulating materials include glasswool (fiberglass) and rockwool batts, cellulose fiber, reflective foils and polystyrene foam. Glasswool and rockwool do not burn whereas polystyrene does. Cellulose fiber insulation is often made from recycled newspaper but is mostly treated with chemicals to prevent it from burning too.

Many of these insulating materials trap air. Air is a good insulator and it helps prevent heat from passing through a material.

Reflective foils work differently as they reflect heat either away from a house or back into a room. Insulation can also act as soundproofing, and different materials have different abilities to reduce the intensity of sound passing through.

» DID YOU KNOW?

Insulating materials are given an "R" rating, which is a measure of thermal resistance. The higher the R value the better the resistance to heat flow. Higher R value materials should be used in areas that experience severe weather such as excessive heat or cold.

E — External influences

There are two aspects of external influences: integrated gardens, and house and garden structures, such as verandas, trellises and pergolas.

Few architects and designers consider integrated gardens, which can be effective in moderating house temperatures. Many people may not realize that gardens can be designed to take an active role in the heating and cooling of the house.

For example, deciduous trees and tall shrubs should be planted to allow winter light to pass into rooms while shading the house during the summer. Remember that protecting walls from direct heat radiation will lower the overall temperature of the house.

The positioning of shrubs and trees is important for the success of this strategy. The low winter Sun can cast long shadows while in summer, the Sun, which is higher in the sky, produces shorter shadows. You will need to consider the height and spread that trees usually attain to know where to plant them.

Shrubs and trees can also perform other functions. Plants can be positioned to deflect cold or hot winds from the house by acting as screens and windbreaks. Creepers and climbers can grow on a frame or the wall itself on the western side. Plants can absorb heat and prevent heat radiation from reaching the wall, keeping that side of the house cool.

Plant deciduous trees on the Sun-facing side of the house.

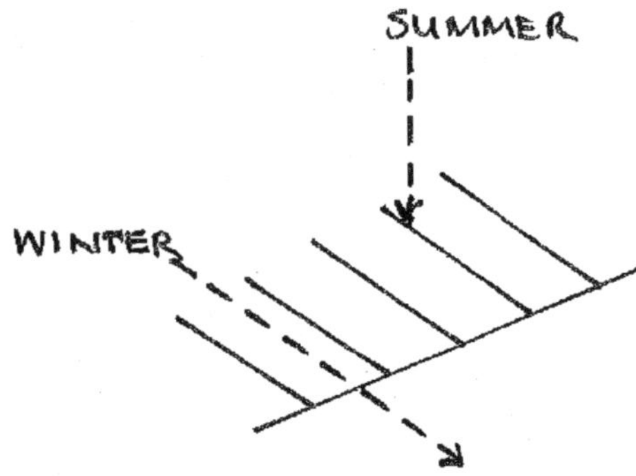

Metal or wood blades, spaced apart at an angle, can shield summer sunlight or direct winter sunlight into the house.

The nature of the tree foliage may also be important. Light-colored, shiny leaves will reflect light onto house walls, adding heat during winter, while dark leaves tend to absorb heat, rather than reflecting it onto surfaces.

Garden structures can also be used for temperature control. A hothouse or greenhouse attached to the sun-facing wall can provide winter heating for the house as well as being used to grow food. Vents will control the direction of heat. In summer, the hothouse roof should be vented at all times to limit unwanted heat transfer to the inside of the house. In winter, other vents can be opened to direct warm air into the home.

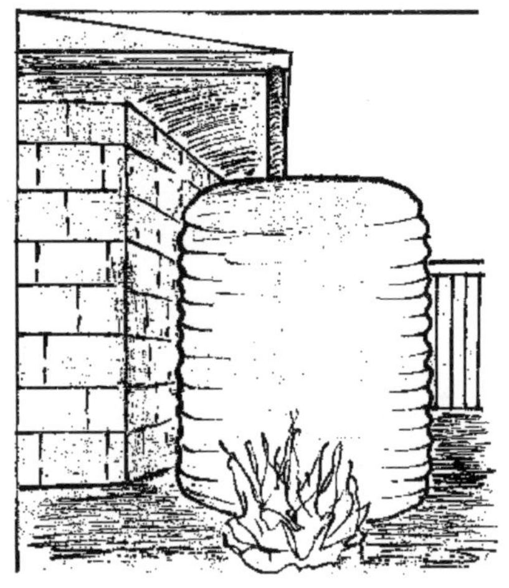

Water has a high heat capacity and can store energy.

Similarly, a shadehouse on the opposite side of the house helps with cooling in summer. Breezes passing through a moist shadehouse on that side of a house can be cooled. This cooler air can be vented throughout the house by opening and closing windows and doors. Any air blowing across pools or ponds is usually cooled in this way before it reaches the house.

Verandas, pergolas and trellises are common garden and house structures that are used to shade walls. Horizontal trellises and pergolas should be placed on the sun-facing and western sides of a house to increase the summer shading of walls. Vertical trellises are more appropriate along western walls.

One variation of the standard pergola is the sun-controlled pergola. This is a structure that has blades set at a particular angle so that only winter sunlight can pass through.

The blades of the sun pergola can be made of metal or timber. With the correct angle and spacing of the blades you will be able to get the maximum winter sunlight to enter the house while completely shielding the roof during summer.

Trellises can either be covered with deciduous vines or shadecloth for summer shading. Being able to remove or retract the shadecloth in winter will allow additional winter heating of house walls.

Even a rainwater tank attached to a wall also helps moderate the heating of a house. The tank should be able to absorb winter sunlight but be completely shaded in summer.

Water can hold lots of heat, which can be slowly radiated and transferred through walls into rooms.

Integrating our garden with the house is an important concept that is often underrated as a means to moderate the effect of temperature and climate extremes.

» DID YOU KNOW?

Three main pigments in the leaves of deciduous trees become the colors of autumn. Carotenoids are responsible for the yellow hues of the season as well as the coloration of carrots, corn, daffodils, and bananas.

Anthocyanins are responsible for the red, scarlet, blue and purple hues in nature. They comprise the characteristic colors of cranberries, blueberries, red apples, cherries, strawberries and plums.

Tannins are the third prominent pigment, and these cause the brown tones to the leaves, especially those of the oak and elm. After the carotenoids and anthocyanin have decomposed, the tannins alone remain, giving the dead leaves their characteristic brown color.

W — Window placement

The size and placement of glass doors and windows is an important passive solar design principle. Heat is easily lost or gained through glass or by the action of winds.

In a climate-sensible house, windows are placed on particular sides of the house to exploit cooling breezes and to facilitate good cross-ventilation.

Generally, rooms on the west and east sides should have the minimum number of windows, and these should also be small or shaded by trees and verandas.

Sometimes about half of a sun-facing wall could be glass to let winter light into the room. As long as the eave overhang is sufficient the summer sun will be

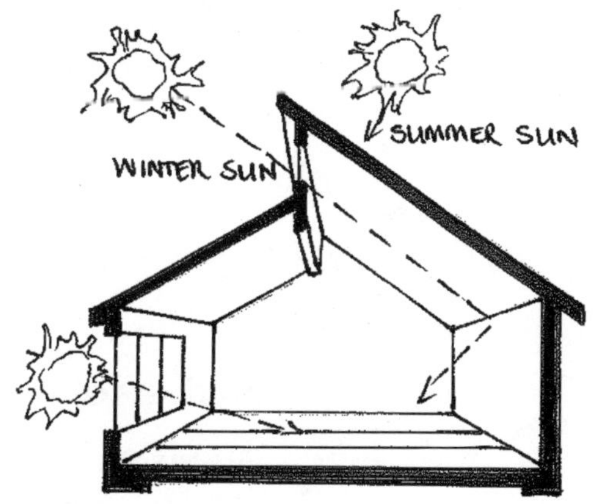

Correctly placed windows will allow additional sunlight into rooms.

screened out. However, too much glass may result in excessive heat loss in winter and too little may mean not enough sunlight is absorbed into rooms.

Sides of the house that are always shaded (south in southern hemisphere, north in northern hemisphere) will tend to be the coldest. Here, double glazing may be helpful. Double glazing (two layers of glass) is common in colder areas of the world as this reduces heat loss from the house.

Window protection is also important to prevent heat transfer. Curtains, reflective foil, tinting, blinds, awnings and shutters can be used effectively to shade glass.

When nighttime approaches, or during hot summer days, curtains and protective covers can be drawn to reduce heat loss or sunlight penetration.

Everyone can make a difference. Everyone has the opportunity to make energy conservation a part of their daily lives. Remember, it is not more expensive to build an energy efficient home, but it does take careful planning.

It is always better and cheaper to make changes to house plans prior to building than to retrofit an existing home to make it energy efficient.

» DID YOU KNOW?

In winter, heat is easily lost in poorly designed homes. Heat loss due to air leaks is about 20%, through walls and ceilings about 50%, and floors and windows 30%. On a typical house, ceiling insulation can save about 25% of heating costs, and wall insulation a further 14%.

Water and space heating are usually the largest consumers of energy in a house, accounting for about 27% and 42% respectively of an average household energy bill. Consequently, the type of water and space heating used in a dwelling has a considerable influence on energy costs and associated greenhouse gas emissions. If Australians were to cut their greenhouse gas contribution by only 1%, then greenhouse gas emissions would be reduced by 1 million tonnes each year.

Active solar systems

Active solar systems have to use pumps to move air or water, or power to operate motors and controllers. While the sun is involved in heating the water in a solar hot water system, for example, energy is used by a pump to transfer water to a storage tank, which is often on the ground.

A simple heat exchange system where a hot coil sits inside a cool water tank is a good way to link a low pressure heating system (such as the solar water panels in the diagram opposite or from the water jacket in a slow combustion stove) to a high pressure house supply system.

Some solar electricity panels (photovoltaics) are mounted as an array and can move to "track" the sun. While some of these devices rely on passive heating of gases to move the array, other devices use small amounts of electricity to drive a motor to turn the panels.

Even though energy is used in active systems, the advantages of higher efficiency and greater production far outweighs the cost of the power consumed.

One or two solar air panels mounted on a roof can provide supplementary heating of the house in winter. Temperature sensors in the house and on the air panel are connected to a controller. When the house temperature is say 20 degrees cooler than the panel, a fan is activated to draw cold air from the house, blow it across the panel where it heats up by the sun and then back into the house. The system is turned off during summer.

Heat pump hot water systems are similar in operation to reverse-cycle air conditioners. They use a fan to blow air over an evaporator coil pipework that contains a refrigerant. The liquid refrigerant changes into a gas, which is then pumped through a compressor to increase the gas's temperature (and pressure).

The excessive heat of the gas transfers through a heat exchange pipe system to heat the

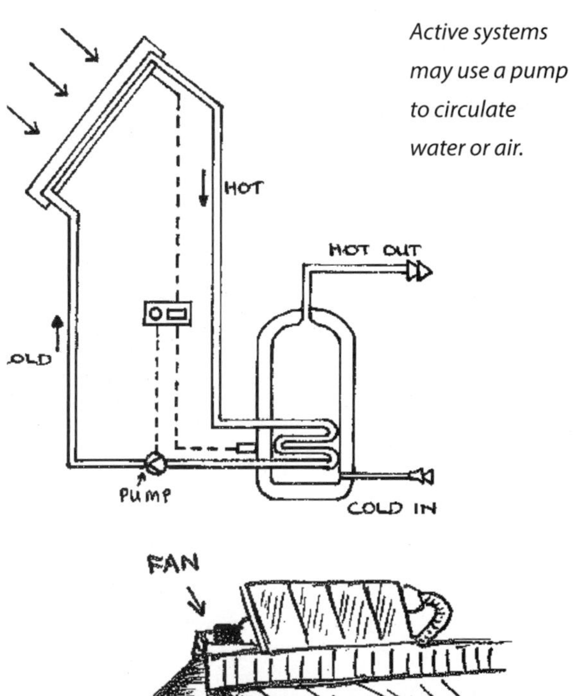

Active systems may use a pump to circulate water or air.

A solar air panel mounted on the roof of a house.

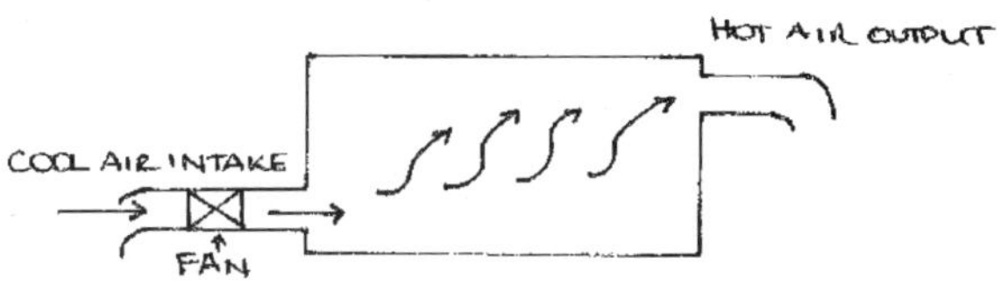

A solar air panel can produce cheap heating of rooms in winter.

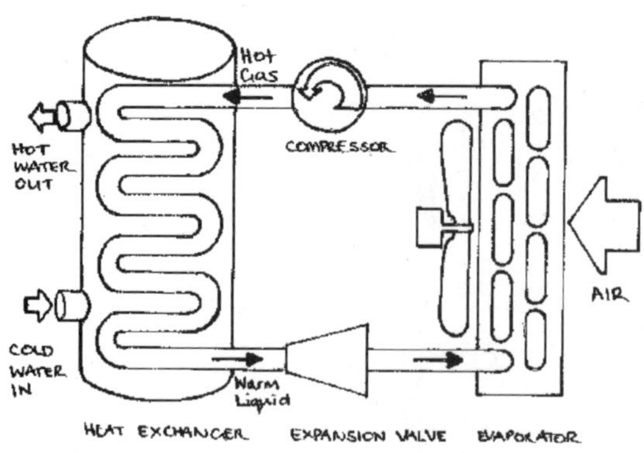

A heat pump extracts energy from the air to heat water.

surrounding water, and the gas subsequently cools back into a liquid. The cooled liquid flows back to the evaporator coil and fan area and the cycle is repeated.

The advantage of a heat pump is that it simply uses the energy of the air to heat water, and it produces more heat than the energy used to achieve this. If you can preheat the water (via roof-mounted solar water panels) before it enters the heat pump storage tank, then even less energy is used to operate the fan.

» DID YOU KNOW?

The ratio of energy produced to that used to make it is known as the coefficient of performance. An electric storage hot water system uses electricity to heat water, and at best the coefficient is about 1 (but usually less). Heat pumps can have a coefficient of performance between 2 and 5, which is a significant heat (energy) saving.

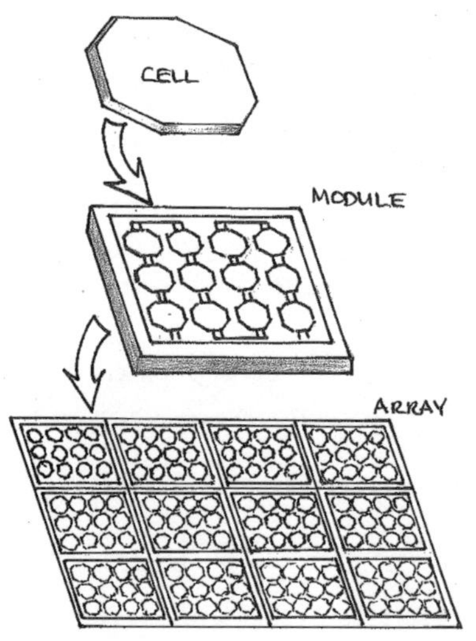

Many individual cells make a module and many modules are joined to make an array.

Energy generation

There are two aspects to energy generation. Primarily, it is about producing electricity, but it also may involve using energy to provide heat and power, move water or be burned as fuel.

Electricity generation

The use of renewable energy to provide electricity and power has been discussed in reasonable detail in my *Basics of Permaculture Design* book, so only a brief summary of photovoltaic, wind and hydro systems are detailed here.

Photovoltaic cells

Solar electricity panels (solar cells) are getting more efficient and cheaper as years go by. As sunlight hits the surfaces of particular elements, electrons from the atoms are released and these flow as electricity (hence the name photovoltaic — photo = light and voltaic = volt, electricity).

Individual solar cells are mounted as a module or panel and many of these are joined to form an array.

The cells are based on a sheet of the semiconductor silicon, most often dosed with small amounts of other substances, which may contain gallium, arsenic, cadmium and germanium.

Emerging research has seen the development of thin film solar cells, including some that incorporate light-sensitive dyes and organic polymers, which have been used to create walls of buildings and even solar cell windows.

Other thin film cells, such as the copper indium gallium selenide cell and the gallium arsenide cell, have demonstrated the most efficiency of converting light into electricity.

Most solar cells installed around the world only have efficiencies of 6–20%, but some new technologies (called triple junction cells) have shown over 40% efficiency.

Solar paint is also being developed where semiconductor polymers are suspended in water and then coated onto plastic or glass sheeting.

Eventually, you may be able to simply paint the solution onto a roof to generate electricity.

Wind generators

Very few houses have access to good wind and a small domestic turbine produces only meager power due to inefficiency and size. Wind is too variable so there are only some sites suitable for wind power. Certainly wind generators can be used in conjunction with solar or hydro systems to complement these.

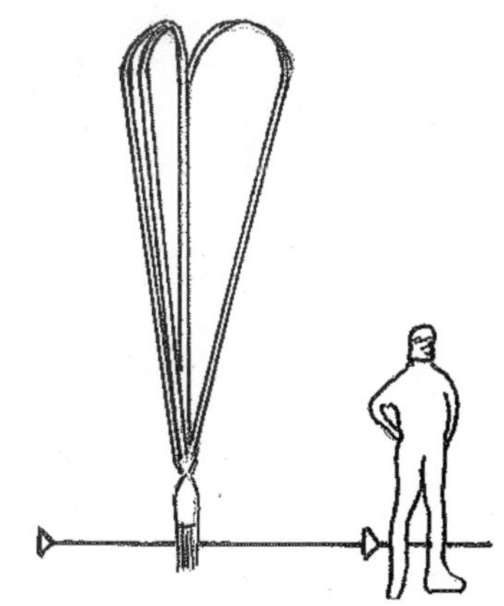

Much research has been undertaken, including this generator design by Ben Storan.

» **DID YOU KNOW?**

The Sun provides about 100W/ft^2 at the Earth's surface. This is enough power to light fifty 20W compact fluorescent light bulbs.

Silent turbines are not yet available in the marketplace so you need to be mindful of noise — for your family and your neighbors who may not appreciate your energy harvesting

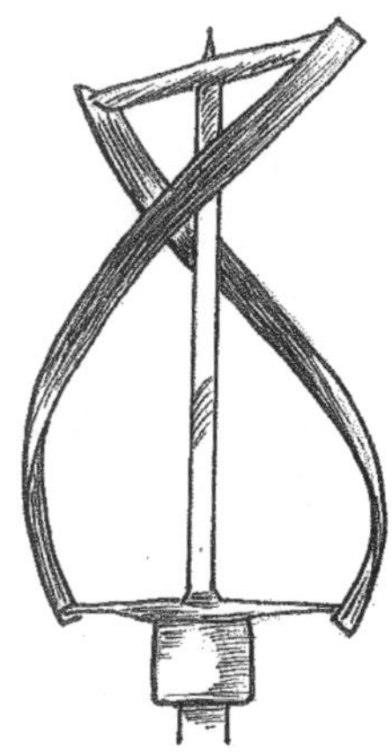

X-wind. Just another example of the many innovative wind generators now becoming available.

endeavors. Generally, to generate reasonable power you need larger blades and taller towers.

In recent years turbine technology has changed and weird and wonderful designs are being built. These new wind generators are more efficient and respond to lower air speeds than the traditional blades, which have been used for decades.

» DID YOU KNOW?

Householders who generate power from photovoltaics or wind generators, or some combination of these, have two options. Systems can be grid-connected where your domestic or commercial system is interconnected to the state service provider. You sell and supply the power you generate to the "grid".

Remote homes may not be able to obtain electricity from a government or private energy supplier. They require a stand-alone system to produce enough electricity to run their household or property. This is often called a RAPS — Remote Area Power Supply.

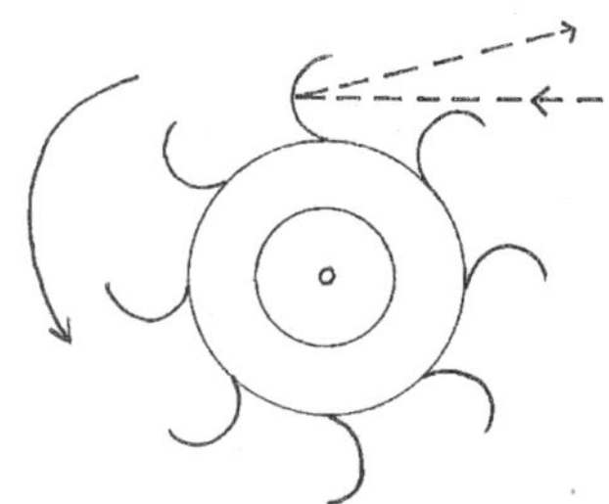

An impulse turbine spins when water is directed onto blades.

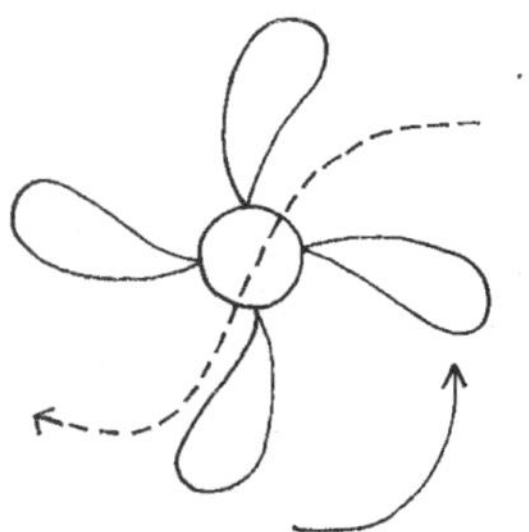

Reaction turbines spin as water (or air) passes through them.

Energy from water

Moving water has inherent kinetic energy. This can be harnessed and converted to mechanical or electrical energy and then used to operate machinery or power homes. However, the energy in moving water or from a waterfall has to be controlled and contained before it can become useful. This may involve pipes from the high water point to a lower point, some weir, dam or storage of water at the high point and suitable equipment, like a turbine, to convert the energy in moving water into a form that can be better used.

Turbines are grouped into two types: impulse turbines and reaction turbines. Impulse turbines spin when a fine jet of water is directed onto the cups or blades. Pelton wheel, cross flow turbines and water wheels are common examples of these devices. Under-shot and over-shot water wheels have been around for literally thousands of years, while most of the other turbines were developed less than 200 years ago.

Reaction turbines don't change the direction of water flow as much as impulse turbines. Reaction turbines simply spin as the water or air passes through them.

A windmill is a very common example of a turbine that reacts to moving air, while Francis and Tyson turbines are common examples of water turbines.

Some of the water reaction turbines can sit in the stream itself and be suspended by a floating raft or anchored to the bank of the stream.

The amount of power that turbines generate depends on the vertical "head" of pressure and the flow. Fast water movement produces more spinning action and more energy can be transferred from water that is dropped from a great height.

Every turbine has particular operating parameters: some work well in high flow situations, others require high head and low flow and others again maintain good efficiency rates when there are variable flows.

Mini hydro systems are only suitable in properties that have a continuous running water supply and a slope over the land.

Unless you have a steam or creek that flows all year round, with a vertical drop of at least a few yards, generating power from water may not be an option for you. It may be better to examine the possibility of solar or wind power.

A hydraulic ram can lift about 10–20% of the water that passes through it to great heights.

Besides generating power from running water, air-driven pumps, called hydraulic rams, can be employed to move water from a stream to a header tank or dam. Hydraulic rams are simple devices that capitalize on the energy in water to compress air in a pump mechanism to lift water.

Pedal power

Both wind and water generated electricity requires a turbine or generator to convert moving energy into electricity. Many bicycles are fitted with a small generator attached to a wheel that spins and makes enough power to light up a headlight while riding at night. Some manufacturers have adapted this principle to make a range of human-powered machines to enable

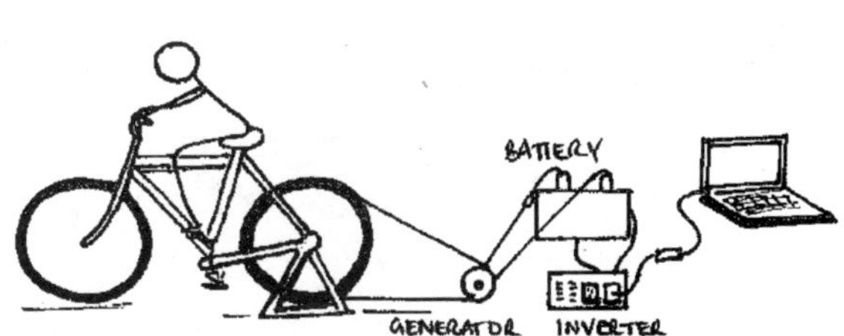

Disused bicycles can be converted into electricity generators.

batteries to be recharged, lights to operate and to drive a water pump and a number of household appliances.

You can even buy a "kit" to convert an unused bicycle into a generator, and use this to both power the appliances and exercise at the same time. For example, a laptop usually draws about 50W, so a 20 minute exercise on a bicycle generator will power the laptop for one hour.

Many bicycle generators can produce up to 500W, so this is enough power to enable an inverter to operate simple household appliances such as kettles, toasters and blenders.

» DID YOU KNOW?

Many homes have appliances that have a "standby" mode. Even when you think that you are saving power, these appliances, such as computers, TVs, stereos, DVD and CD players, contribute to about 5% of greenhouse gas emissions and cost you about $100 each year in power bills.

Biological fuels

Biogas

There are a variety of gases useful as fuel, and these include liquefied petroleum gas (LPG), natural gas and biogas.

Homemade biogas generators are common in some countries.

LPG is a mixture of volatile fractions from petroleum refining: principally propane and butane, with small amounts of propylene and butylene.

This is often used as a substitute for gas in motor vehicles because it is easily liquefied and has a reasonably high fuel (or calorific) value when burned.

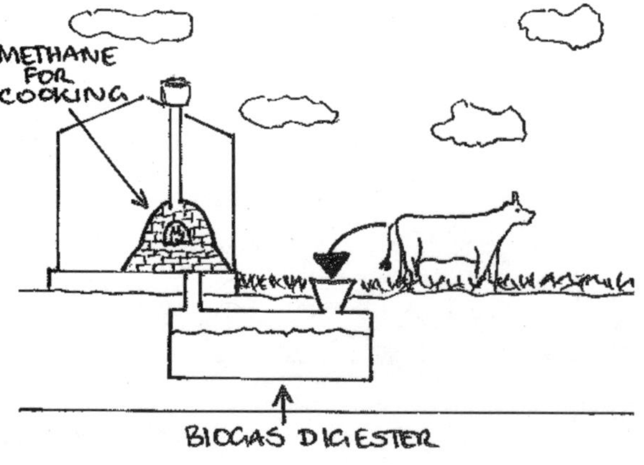

Natural gas is a fossil fuel typically trapped and buried under ground or under the seafloor. It is mainly methane, but does contain smaller amounts of other organic substances as well as carbon dioxide and hydrogen sulfide. Natural gas is harder to capture and store as it is a gas at room temperature.

Biogas is a by-product of the anaerobic decomposition or fermentation of organic matter and has a similar chemical composition to natural gas. It is easily made from a few simple substances and inexpensive equipment. Biogas can be produced from many organic substances, such as food wastes, plant residue, manures, fruit cannery wastes and landfill green wastes.

Biogas can be used as a fuel for heating, cooking and steam generation, or as a fuel in internal combustion engines. It burns gently and produces carbon dioxide and water, with little or no poisonous carbon monoxide, so it is relatively safe as a fuel in a home.

When biogas is made, it may contain reasonable levels of hydrogen sulfide and carbon dioxide, and these need removing in an operation called "scrubbing."

The main purpose of scrubbing is to reduce these corrosive gases, which combine with the water vapor to form acids and hence corrode all metal parts of the gas system, as well as to get rid of the unburnable carbon dioxide that simply "takes up space" for no useful return.

To remove these gases, the biogas is bubbled through water and then passed through steel wool, iron shavings or something similar to remove the hydrogen sulfide.

In an optimum operation you can produce about 9–12 ft^3 of gas (250–300l) in a 50 ft^3 digester, which has about 4% solids — that's 130 lb of wastes in approximately 400 gal (1,500l) of water.

This volume depends on the temperature of the system and the type of solids added, so returns are mostly lower than this. At a practical level, one bucket of manure is typically required to produce enough biogas to operate a stove for the day's meals.

Cows produce 65 lb/day — about two bucketsful, while pigs produce about 13 lb/day and chickens much less than 1 lb/day. (One cow produces over 1 ton dry matter each year, which is equivalent to about 10 tons of fresh manure.)

Generally, humans do not produce enough solid waste (even with daily food scraps thrown in) to produce enough biogas to run a stove. Human waste only makes about one-quarter of the biogas produced from the equivalent amount of cow manure.

Some research is being undertaken with algae and fungi to produce biogas. Various algae and microalgae are being cultivated in oceans to provide the feedstock to be used in biofuel (biodiesel) production or fermented directly to make biogas.

» DID YOU KNOW?

The average household in Australia throws out about 30 lb of waste each week. Half of this is food scraps and garden refuse, which could easily be composted. Composting your organic waste would save about 900 lb of landfill each year.

Fungi are being used as a pre-treatment strategy. Fungi change and digest cellulose and lignified material, such as hay and straw, to enable microbes to ferment these materials easier, resulting in greater biogas production.

Biodiesel

Biodiesel can be a substitute for diesel in engines. Sometimes it is blended with petrodiesel but can be used "neat" with minor engine modifications, or even no modifications at all.

Biodiesel is made by chemically combining fats (such as vegetable oils and animal lard) with an alcohol to produce organic compounds known as esters. These substances are volatile and burn easily, even though the calorific value is slightly less than petroleum-based diesel.

Typical feedstocks (to provide the oil or fat content) include soybean, canola and sunflower oils, although animal fats and grease from pork and poultry, for example, can be used but with diminishing returns. Backyard operators often obtain the cooking oil and fat wastes from fish and chip shops, filter it and then add the reagents.

Plant oils can be used to make biodiesel.

The most common alcohol used for the conversion of oil into esters is methanol (and ethanol is an option), and a catalyst, such as sodium hydroxide (NaOH) or sodium methoxide (CH_3ONa), is used to speed up the process, which may still take several hours.

While the majority of what is produced is biodiesel, many by-products are also formed and these have to be removed. Soap, glycerol, excess alcohol and some water have to be decanted or drawn off, and the relative amounts of each of these depend on what oil or fat and what catalyst was used.

Biological heat production

How to make a hot compost pile was discussed in chapter 4. While you wouldn't want to take all of this heat away, as microbes use the energy to break down the organic wastes, some of it can be used to heat water, animal enclosures or our homes.

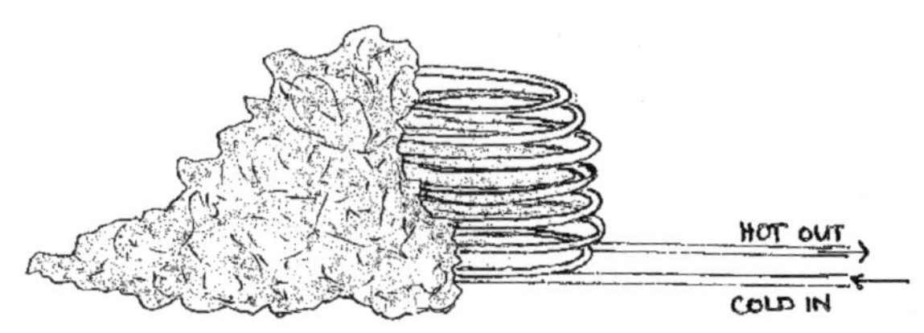

Hot water can be extracted from a compost pile.

As long as the compost pile has the correct ratio of ingredients and is large enough (held within a 70 ft^3 wire cage or bigger), the heat generated can heat a coiled pipe and produce hot water for a few showers.

Alternatively, hot water can be passed in pipes through a heat exchange space heater to heat a room or animal pen to keep everyone and everything nice and warm during the night and in colder months.

Generally, the larger the compost pile, more heat over a longer time can be utilized. If the compost is really "cooking" then temperatures a little over 140°F are common.

While black poly irrigation pipe can be used as the coil, soft and bendable copper tubing is far better to absorb the heat and move the hot water.

Bear in mind that even 0.75 in (20 mm) copper piping only holds 10 oz for every three feet in length, so you may need a very long pipe to have a shower.

Even if you only use 7.5 gal for a shower (say 2.5 gal/minute for a 3-minute shower — that's quick!) you need about 120–150 ft of copper coil, assuming that as water passes through the coil it can quickly heat up.

To overcome this problem, a shorter coil is used to transfer hot water to a small storage tank, typically supported on a stand several feet above the pile, so that gravity can be used to operate the shower.

Three to six feet of "head" is not ideal as this results in just a trickle, so you could pump the hot water if that was feasible.

Energy for cooking

Cooking food requires heat or fire, or both. Low technology solutions utilize either solar energy or biological fuels to provide this energy.

A solar oven or solar cooker uses sunlight to cook food. These can be easily made from simple materials and can produce high enough temperatures (often up to 300°F) to cook any sort of food.

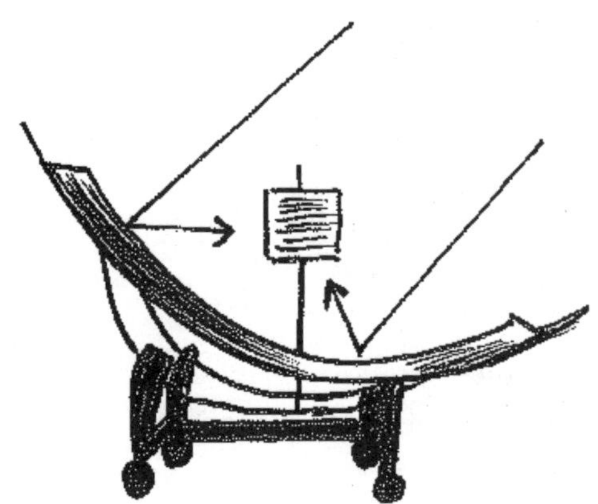

Reflective curved surfaces can direct sunlight onto a pot to cook food.

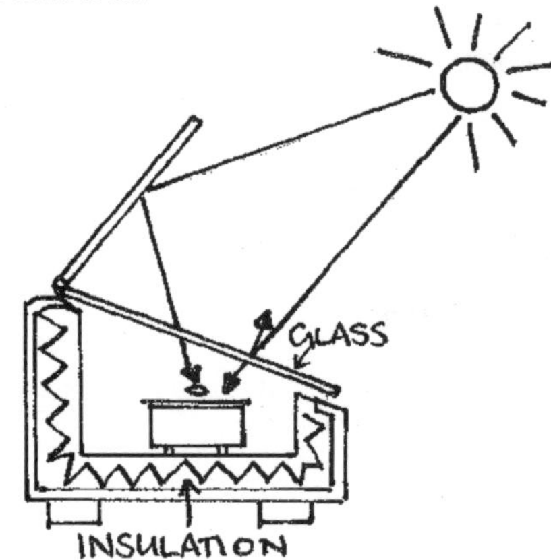

Solar ovens can be made from simple materials.

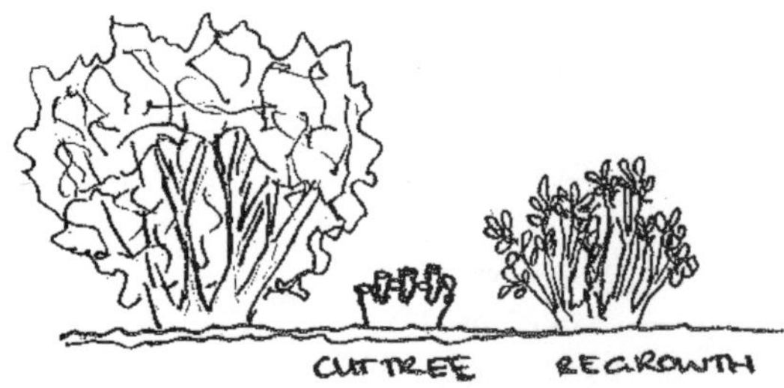

Coppicing fast-growing trees enables a steady supply of wood.

There are two main types of solar cookers — parabolic reflectors and box cookers. Parabolic reflectors incorporate a reflective foil or mirrors to concentrate the sunlight energy to a focal point. The food or pot is held at this focal point and heats up.

Solar box ovens typically consist of an insulated box with dark-painted internal surfaces, and a glass lid. One or two adjustable reflective surfaces may be added to help direct sunlight into the box.

Meals, such as stews, rice dishes, bread, cakes and even roasts, cook over several hours. Occasionally the cooker is moved to follow the Sun's path across the sky to ensure maximum heat gain.

Biological fuels include biogas and wood, but can also include other organic substances such as manure (dung), oil and flammable solvents. Biogas has been discussed earlier, so let's talk about wood.

It is possible to view wood as a renewable energy source. This is because we can grow and harvest wood and timber in a sustainable way. We just need to remove and use wood at the same rate at which it grows, or a lower rate.

Planting specific species of plants that can be coppiced to produce a continual supply of timber and fuel is easy to do. There are many plants that are currently used for these purposes, and others that tend to be fast growing and produce high heat energy when burned can be tried.

Even compressed sawdust, a by-product from the timber milling industry made into pellets, or paper pulp bricks are far less costly than using kerosene, oil or gas for heating and cooking.

The majority of people in the world who use biomass as a fuel burn animal dung. Cow and other animal patties are dried in the sun and burned in stoves. This is not as efficient as burning wood, but in many places wood is in short supply.

Rocket stoves are becoming popular in some countries, as these require smaller volumes of wood to produce heat at a very efficient rate. The combustion of the wood is thorough as the stove has good air flow and the generated heat is better utilized and not lost "up the chimney."

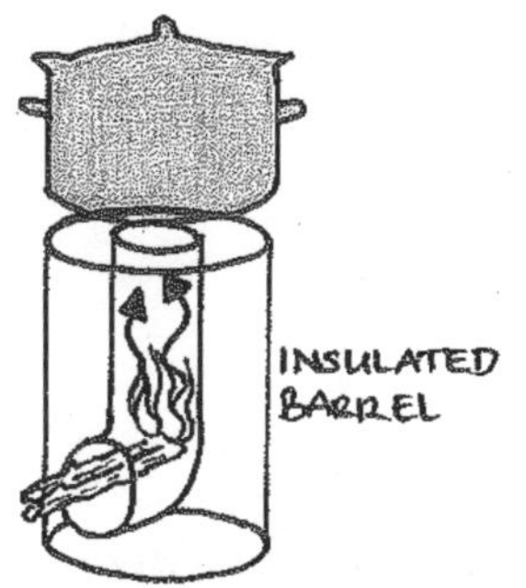

A rocket stove is very efficient as the generated heat is retained.

A rocket mass heater is a recent variation of the rocket stove where the fuel burns sideways, the exhaust gases are relatively cool (and clean) and there is hardly any smoke.

Strong convection currents inside the insulated riser tube draw the flame into the heater. The majority of the heat produced stays inside the heater and is not lost as it would be in a conventional wood fire.

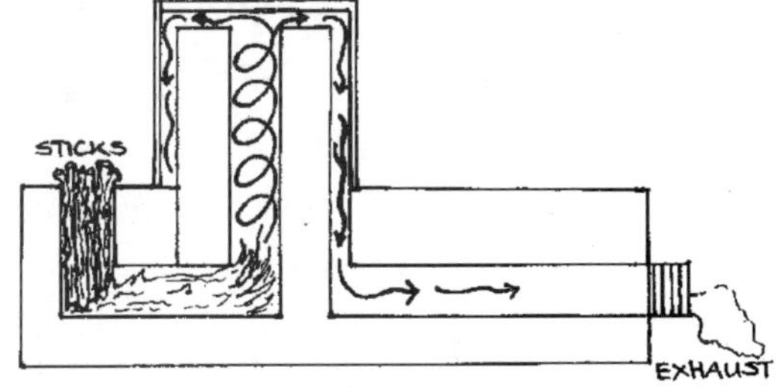

A rocket mass heater produces heat from relatively small volumes of wood or fuel.

Wood-fired ovens are also becoming popular in outdoor kitchens. These "pizza" ovens are easy to make but you can buy readymade kits and ovens themselves. Most use refractory cement to prevent cracking but many people who build "do-it-yourself" ovens just use a clay and sand (or clay and straw as in a cob oven) dome built on a brick floor. Normal household bricks can be used but fire bricks are best. Refractory cement is a purpose-made product and makes a fire-resistant mortar for use in kilns and furnaces.

Pizza ovens are becoming popular outdoor kitchens.

The most popular mortar for a pizza oven is a hybrid mix of refractory cement, normal cement, sand and lime. This allows the surfaces and joints of the oven to be sealed like a ceramic. Fire bricks have a high alumina content and this ensures they can withstand very high temperatures without cracking and falling apart.

Energy for cooling

Most people probably think of air conditioners when they investigate ways of cooling the rooms of their homes. Cheaper and lower operating cost options include ceiling fans, evaporative coolers and whole-house fans in the roof space.

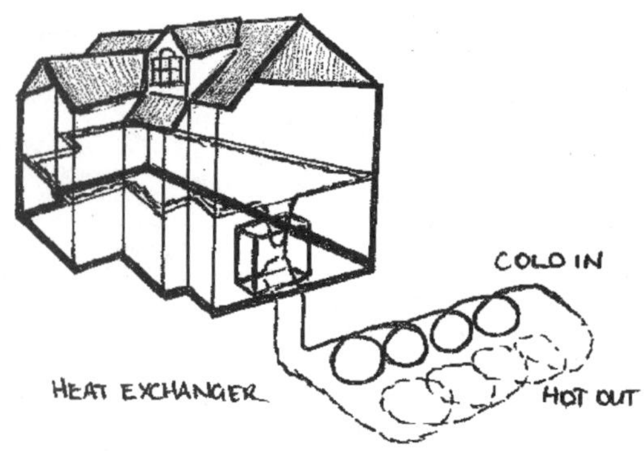

In some countries, water piping buried in the ground can be cooled or even heated to produce supplementary cooling or heating in a home.

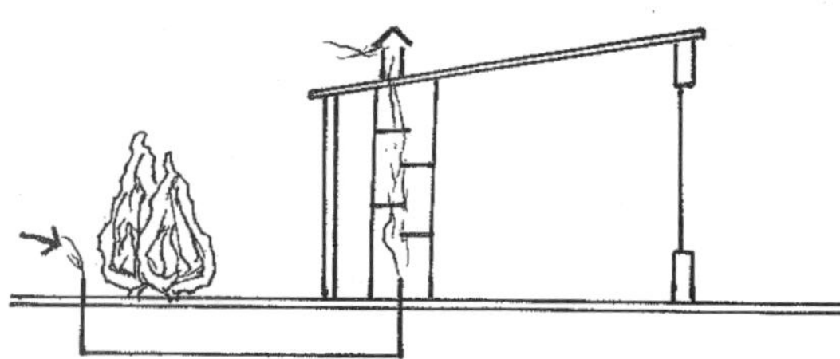

Air can be cooled as it is drawn through a pipe buried underground to keep vegetables fresh.

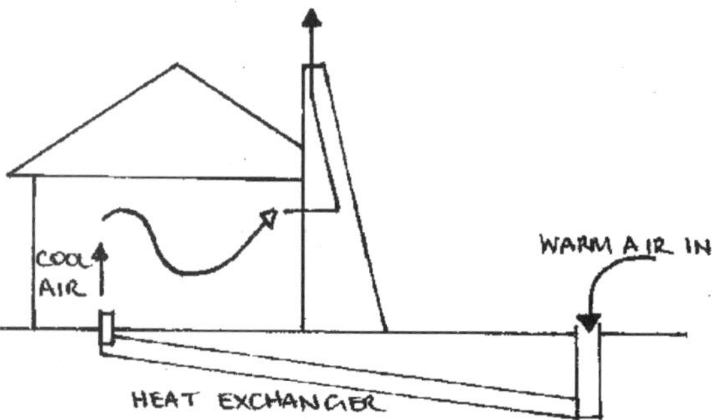

Piped air can also be effective as a simple air-conditioning system.

Reverse-cycle air conditioners are heat pumps that can be used to both heat and cool a home, depending on the ambient temperature, and tend to be more efficient and have cheaper operating costs than whole-house ducted air-conditioning systems.

If houses are designed to be passive solar, then a ceiling or freestanding fan is all that is generally needed to cool a house down in hot weather spells. Ground-source heat pumps (also called geothermal heat pumps), which have pipes buried under the ground, are one way to further reduce cooling (and heating) costs in a home, and while these are relatively unknown they certainly warrant investigation.

Using the ground to cool a home is not new. Air drawn through pipes buried about a yard or two below the ground can be ducted into rooms or used to cool fruit and vegetables as part of a solar (thermal) chimney.

A black-painted chimney heats up during the day and the hot air rises and is ducted out of the house. This causes a suction of air along the pipe. This air is cooled by the earth and enters the house as much cooler air.

Passing this air through an insulated cupboard keeps food fresh, while a pipe network under the house enables cool air to enter each room as required, by opening and closing a vent.

Evaporating water can also cool air. The heat of the air can be used to evaporate water and while this is the principle of evaporative air coolers, a device called a Coolgardie safe is used to keep food cooler and longer before any spoilage occurs.

Burlap, with one end dipped into a tray of water on top of a cupboard and allowed to hang down, absorbs water and holds it in within its fibers.

As a breeze passes over the burlap, water evaporates and the air passing through the bag is cooled. For water to change from a liquid to a gas, it requires energy.

The heat in the air (of the breeze) is enough to enable the change of state of water, and so the air cools.

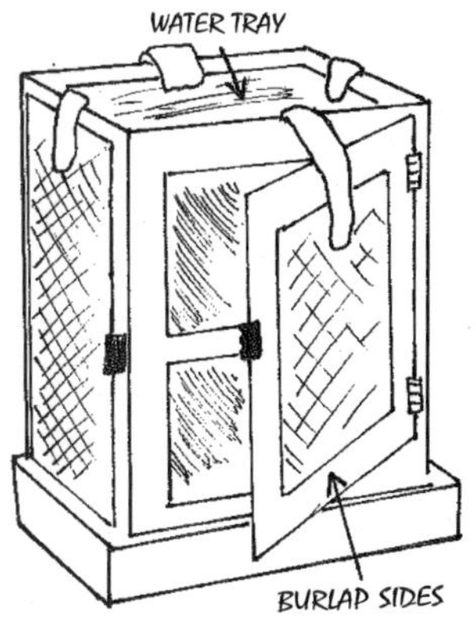

A Coolgardie safe cools air as water evaporates from its sides.

Energy for heating

Besides burning fuel, using the heat from compost production and employing a heat pump device, solar energy can be used for heating. Sunlight can be harnessed to heat water, dry foods and produce freshwater.

Solar water heaters are devices that capture sunlight, convert this to heat and then use the heat energy to produce hot water. Some systems have the storage tank on the ground (active systems), others above the solar heating panels (passive systems). In a passive system heated water rises towards the horizontal storage tank and colder water falls downwards into the collectors.

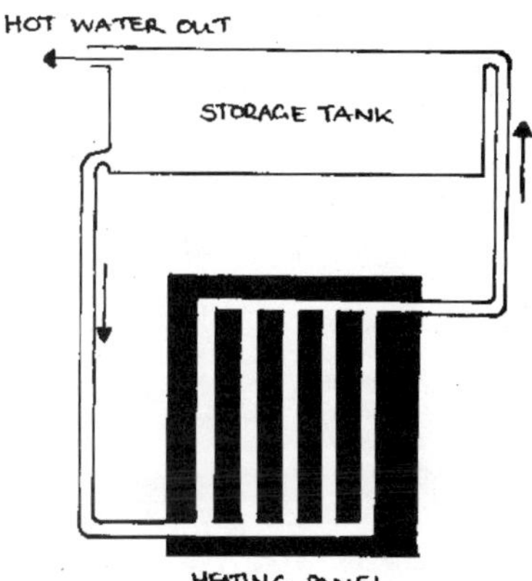

A passive solar hot water system utilizes natural flows of warm and cold water to make hot water.

Thermosiphoning occurs and there is a steady flow of water into and out of the storage tank. On low sunlight days or during continuous cloudy weather, the water temperature may remain low. A booster is required to heat the water enough for its use inside the home. The most common boosters either use electric heating elements or gas burners.

Active solar hot water systems use pumps to move water from a storage tank on the ground to the roof area where the collectors are mounted. They may still need a booster, but they tend to lose less heat and be more efficient than passive systems.

A solar still is also used to heat water but it enables the production of pure water from salty or contaminated water. Any type of water source can be used, and the sun's energy causes some of the water to evaporate, leaving the salts and pollutants behind. The water vapor is trapped and condensed, so that pure

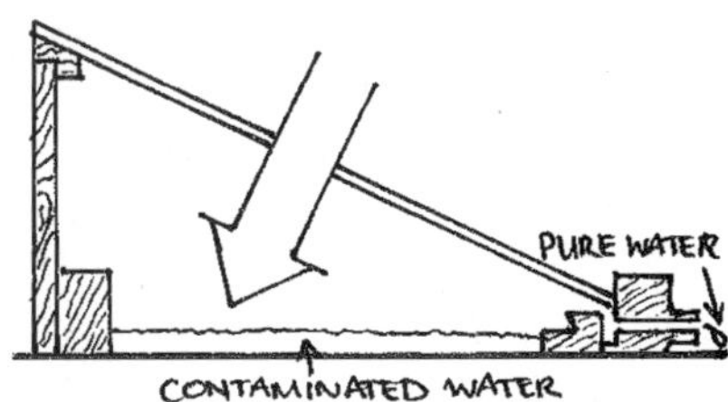

A solar still makes small amounts of pure water by only using sunlight.

A homemade fruit dryer uses clear plastic or glass sides and top to desiccate fruit.

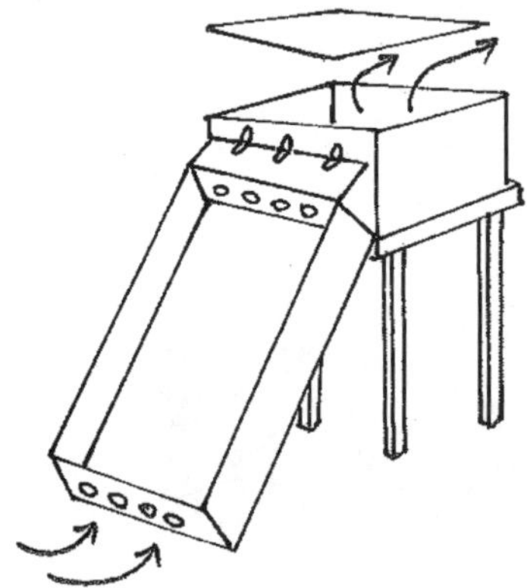

Sunlight heats air, which passes over and through shelves containing fruit slices.

Sometimes, a simple tool such as a scythe saves us time and energy.

liquid water is able to be collected. A solar still generally doesn't provide huge volumes of freshwater each day, but certainly enough for survival and emergency water.

The other advantage of a solar still is that it can heat water to a high enough temperature to sterilize it. Most microscopic organisms die at 140 to 160°F, and this temperature is easily achieved in a solar still.

Solar dryers are typically used to dry fruit, as a method of food preservation. Sliced fruit, meat and herbs generally take a day or two to dry. Not all of the moisture is lost, so fruit tends to be leathery and a little flexible.

There are a few different versions of solar dryers, but they rely on the heat of the sun to cause air to heat up and to move upwards and through the trays, slowly drying the fruit.

Several trays, often with window screen mesh as the base, allow air to move through and over the thinly-sliced fruit and vegetables.

Simple technologies

I am a firm believer that when you have a machine, or can afford to hire one, use it. You wouldn't spend days or even weeks digging holes, shifting soil for major landscaping works and creating roads and dams all by hand. Machines save us time and work, even when many consume fuel.

However, there is always opportunity for people power. If you can't afford machinery then manual labor will have to do. One day maybe all vehicles will run on biodiesel made from farm waste.

Nevertheless, there are times when a scythe is more appropriate than a string trimmer, a horse and plow better than a large tractor, hand tools instead of power tools, and an ax less hassle than a chainsaw.

Unfortunately many people will need to re-skill or learn new skills if they ever need to use simple tools and techniques, but there is nothing wrong with that! As we all move into a new era of human history, let's embrace the consideration and use of appropriate technologies.

FORGOTTEN SKILLS

16

SPINNING WOOL, macramé, knitting, crochet and dressmaking. Once common, probably up until the late 1960s, and even taught in schools, these are many of the skills performed by our mothers and grandmothers that have been lost in the current generation or two.

I remember as a boy we had, in our suburb, a baker (actually two), a bootmaker, an iron foundry, corner store, local garage and mechanic (where I worked during some school holidays actually serving petrol to customers and doing basic car services), a fruit and veggie man that drove from house to house, as well as the usual doctor, dentist, hairdresser, police station and local post office. How things have changed.

I also had a holiday job at one of the bakeries that specialized in pastries. I learned how to make pies, pasties, cinnamon buns and other cakes and goodies.

Most teenagers these days work for a fast food outlet or are "checkout chicks" (and I include boys in this too). Teenagers rarely learn how to prepare and make real food. I learned how to wash, peel and cut potatoes, but most fast food employees simply open the packet of frozen precooked food. And while schools do give children a taste of some of these skills, it just isn't to the same depth of training young people for employment as it once was.

You might think that in my day where girls at school learned how to cook, sew and knit and boys learned woodwork and metalwork that it all seems a little sexist. But learning these skills early in my life gave me the confidence and interest to build houses, repair structures and be a little inventive — skills I can now pass on during my permaculture courses.

I love all of my grandchildren, but all many of them seem to do is play videogames, watch TV, "text" messages on their iPhones, and be so absorbed in social media that they seldom climb a tree, go for a forest walk, build a playhouse, make a simple raft or canoe and paddle down a local creek, or ride their bikes with their pals over the "jumps" they built down the road on a vacant piece of land. Yes, some play sport at various times of the year, but there can be so much more about being a kid. How times have changed.

We could write about a myriad of "forgotten skills" that more recent generations of people do not do, but here are ten simple things to start you in the right direction.

Some simple knots

There are literally hundreds of different types of knots. A knot is basically used to tie two ropes or objects together. If we join a rope to another rope, we call it a bend, and if we tied a rope to a steel post then we describe the knot as a hitch.

To describe how a knot is tied we will use the following terminology. The "working end" is the end of the rope that is manipulated to make a knot — maybe the shorter piece. The "standing end" is the main length of the rope, maybe the longer piece, often under load.

The following five knots are very useful and easy to learn.

Sheet bend

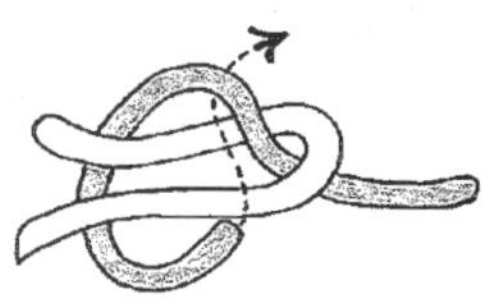

A simple knot to tie two different-sized ropes together. Sometimes you run out of rope and just need a little bit more. You can attach another type of rope of different thickness by using the sheet bend.

1. Form a loop in one end of your main rope (standing end). Pass the end of the other rope (working end) through the loop, around the back of the main rope and then back under itself.
2. Pull the ends of both ropes to tighten.
3. For additional strength, wrap two turns around the main rope (double sheet bend).

Clove hitch

An all-purpose hitch to tie a rope to a post or rail, such as when you tie down a load on a trailer or pickup truck. The knot is easy to undo by a firm jiggle and enough of the knot is loosened to enable untying. If the tension is not maintained then occasionally this knot becomes unravelled.

1. Pass the end of the rope over the rail and then under itself. Pass the rope around the rail again.
2. On the second turn, slip the working end under the last wrap.
3. Pull both ends of the rope tightly.

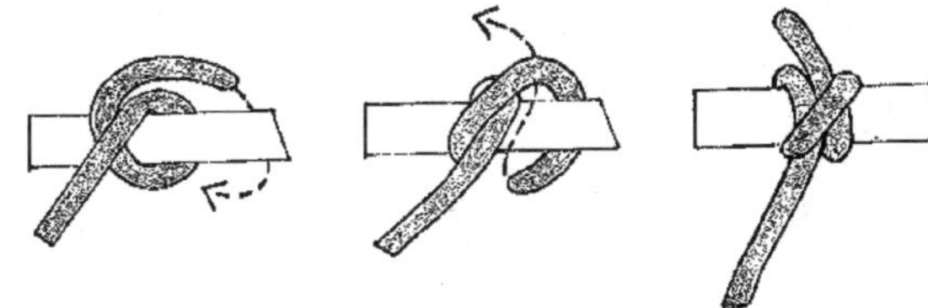

Trucker's hitch

This is the best knot to secure a load on a trailer or truck. It allows you to tighten the rope, ensuring the load stays in place. The knot is then secured by using the clove hitch for the end of the rope.

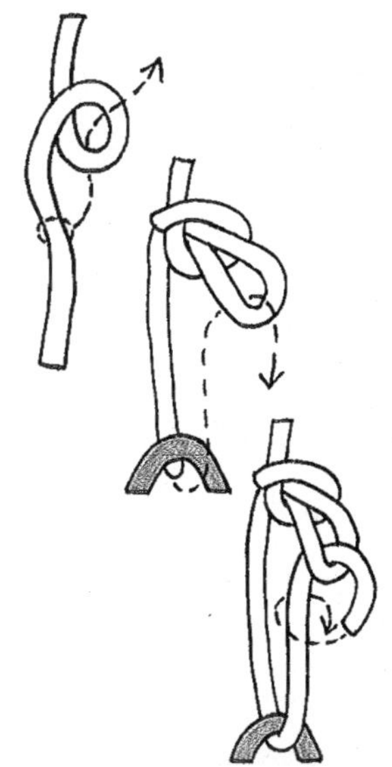

1. Make a small loop about 3–6 ft from the end of the rope. Gather up another small loop and pass this through the first one. Gently pull down to tighten the first loop onto the second.
2. Pass the working end of the rope under and around the rail or post and thread it back through the loop.
3. Pull tightly to drag the standing rope downwards (using the loop as a pulley) to secure the load.
4. Tie off the working end by a clove hitch or other knot to prevent the trucker's knot becoming loose.

Transom knot

This is a lashing knot and ideal for tying two poles together (at right angles) when you are making a bamboo fence or erecting a tent. This can be a permanent knot and the ends of the rope trimmed as required.

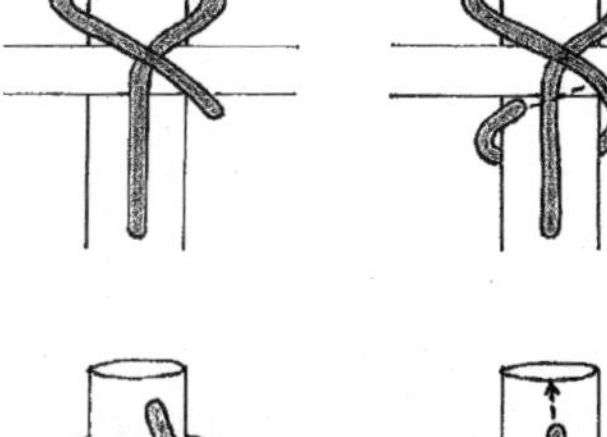

1. Holding the poles at right angles, pass the working end around the back of the rear pole.
2. Pass the rope end across the front pole and then around the back of the rear pole again.
3. Before it becomes too tightened, feed the working end over the standing end but under the second turn.
4. Pull both ends tightly to secure poles together. Trim the rope, either side of the knot.

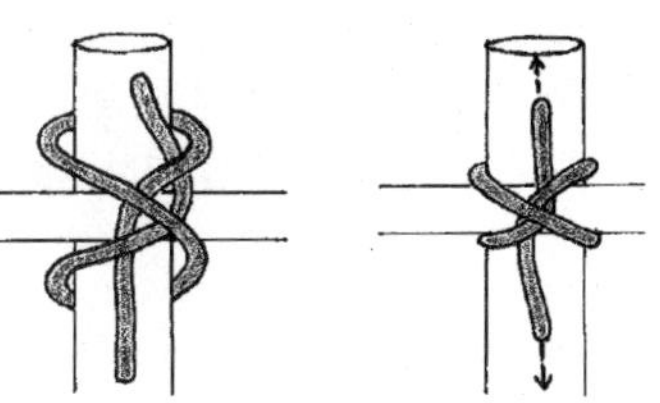

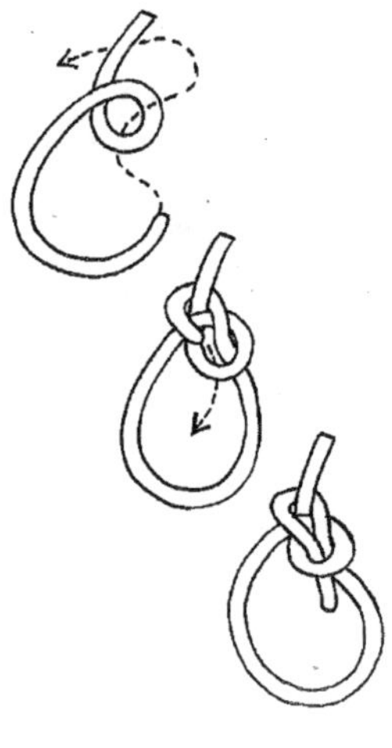

Bowline

The bowline forms a secure loop that can be used to stake and hold a tree upright, tie around a handle of a bucket or tool to lower into a hole, and to place over a person or object to lift them out of a hole or off a cliff face. It is known as a rescue knot.

1. Form a small loop in the working end.
2. Bring the free end up, through the loop, around the back of the standing line and back again through the loop.
3. Tighten the knot by pulling on the free end while holding the standing line (so the new larger loop doesn't slip too much).

Handmade paper

Paper is made from the pulp of trees. In many large cities in the world it has been estimated that somewhere between 3,000–4,000 trees are used each week to produce the weekend newspapers. Trees are rapidly being depleted globally, leading to dehydrated landscapes, soil and wildlife habitat loss and streamline silting. Consumer demand has outstripped tree growth and replacement. Recycling paper, cardboard and tree material will be inevitable in many countries in the years to come.

Old newspaper can be used to make paper but if it contains excess ink, the paper produced will be stained a gray color. Computer or white paper pieces work the best. Glossy paper and magazines should be avoided because they are difficult to mash.

You can mash and shred plant materials such as the leaves and stems from bananas, but these are more time consuming and may require chemicals, boiling and physical mashing to make the pulp. If you want to give it a go, combine about 20–30% of banana pulp with the balance of recycled paper pulp.

The basic principles of this process can be easily demonstrated but you will need a deckle. This is a wooden frame that has a fabric (screen) stretched over it. A typical frame might use 1 in (width and height) timber and be about 5 in^2.

The fabric needs to have holes in it to allow for water drainage. Window screen or even shadecloth is suitable. Secure the window screen or shadecloth by gluing or stapling it to the frame. The deckle should be constructed by keeping both the size of the tank and the size of the paper sheet you want in mind. The deckle should easily slide in and out of the tank or basin.

Beware of using colored felt or cloth. The colors may run and stain your paper. Commercially made paper is often bleached by chemicals that can damage our environment. This exercise uses no bleach or chemicals in the process.

Materials needed

Deckle
Waste paper
Electric blender or food processor
Large tank, pneumatic trough or baby's bath
Felt pieces or cloth (e.g. old plain handkerchief), large enough to cover deckle
Option: electric iron, flat masonite or plywood, weights or bricks

Method

1. Tear waste paper into small pieces and place into a food blender.
2. When the blender becomes half-full with paper add enough water to completely cover it.
3. Blend the paper until no more large lumps are visible, and the mixture is of a thin consistency.
4. Pour the mixture into a large tank or trough.
5. Add an equal volume of water to the tank.
6. Slide the deckle at an angle into the tank. Hand mix the paper fibers in the water (to make an even mixture).
7. Lift the deckle upwards, with a slight sideways movement (or tapping) to filter the fiber and drain the water.
8. Turn the deckle over and place it onto a felt cloth or handkerchief.

A deckle

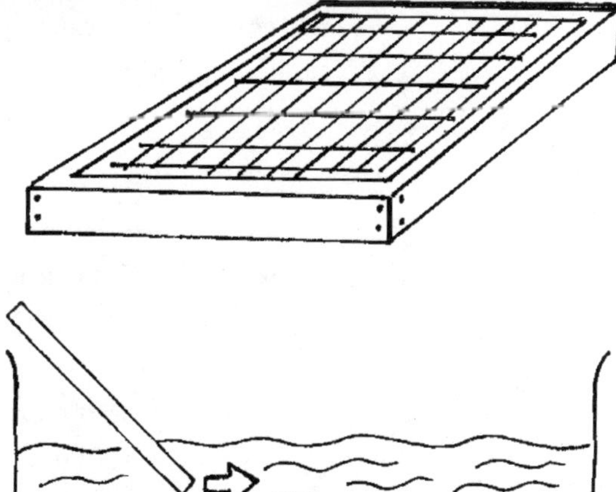

Slide the deckle into the pulp bath.

Lift the paper carefully off the cloth.

9. Wipe the underside (now top) of the deckle to remove excess water. Rub the wire to free the fibers so that the paper can be separated from the frame.
10. Carefully pull the frame off the paper. The paper should remain behind on the felt or cloth.
11. Cover the paper sheet with another piece of cloth to help the paper dry. Alternatively, after a piece of cloth or felt is placed on top of the newly formed paper, place a flat board, with or without weights, to flatten and help dry the sheet.

    The drying process can be speeded up with the use of a hot iron. Do not iron the sheets directly; the paper sheet should be sandwiched between pieces of cloth.
12. Before the paper dries completely, gently pull the sheets off the felt — if left too long the paper sticks to the cloth.

Options

You could also try different screens. For example, compare steel wire with plastic window screen or shadecloth. Experiment by adding food color dye to the water. The paper produced is often patchy in color but still useable.

If you intend to write on your paper with ink, then you may have to "size" the paper. This means that the paper is immersed or covered with substances, such as starch or gelatine, to prevent the bleeding of the ink throughout the paper sheet.

» DID YOU KNOW?

Sizing is how paper is protected, so that inks are not absorbed and spread throughout the paper fibers (bleeding of inks). Particular chemicals are added to the cellulose fiber pulp and act as fillers or protective glazes (like a paint sealer) to strengthen the paper and allow dyes to dry on the paper surface.

Historically, alum has been used, but gelatine (animal) and starch (plant) solutions can be either applied to the paper surface with a brush, or sheets are soaked for a minute in a shallow tray. These are then removed, dried and pressed again.

Making soap

Some people believe that soaps and detergents are essentially the same, and simply think of them as different types of cleaning agents. Other people suggest that soap is a solid while a detergent is a liquid.

While this is true, the main difference between these two substances is that traditionally soaps are made from natural substances while detergents are synthetic.

Most detergents are produced from oil. While soaps and detergents have similar functions, the chemical structure is completely different.

Soap can be made from waste fat or oil. Caustic soda (sodium hydroxide) is added to cooking lard or oil and after a series of steps and reactions soap precipitates out.

There are many recipes for soap. However, they all use sodium hydroxide, which is a corrosive and poisonous chemical. You will need to exercise care when using it. Sodium hydroxide should not be directly heated.

The process of soap making is called saponification. Organic esters of fatty acids are changed into sodium salts (carboxylates) by the addition of sodium hydroxide. Glycerol is a by-product in this reaction.

Soaps and the synthetic detergents are surface active agents, surfactants or wetting agents. All three terms refer to substances that permit water to combine with grease and dirt.

Dirty dishes or clothes are cleaned because one end of the soap molecule is non-polar and dissolves in grease while the other end is polar and will dissolve in water.

Soap flakes will lather in water. When a small sample of soap is added to water in a glass and the contents stirred or shaken vigorously, a lather (bubbles) can be observed.

Detergents characteristically produce a greater lathering action. You should perform this test to show whether you have been successful in your soap making venture.

The addition of calcium, magnesium or ferrous ions produces an insoluble salt with the soap molecules. These ions make the water "hard" as little lathering occurs when they are present.

The familiar "bath tub ring" is caused by the precipitation of these substances with soap.

Ingredients

Sodium hydroxide pellets

Lard (pig fat) or cooking (vegetable) oil. You can have almost any combination of oils, such as sunflower, canola and olive oils. Coconut and palm oils are normally solid at room temperature

Two saucepans — one small to fit inside a larger one. Use steel pots as caustic soda reacts with aluminum

Saturated salt (NaCl) solution (dissolve as much salt as you can in 7 oz of water)

Ice

You will also need a measuring cylinder, hotplate, stick (or immersion) blender (or wooden spoon), and some type of filter (filter paper or tea strainer).

Method

This method only uses small amounts of ingredients. Once you have made soap successfully you can scale up the amounts of each ingredient.

1. Place about 2 oz of lard or 3 tbsp of oil in a small saucepan.
2. Melt the fat or heat the oil in a water bath. Use a larger saucepan for this purpose.
3. Use a water bath to heat the soap mixture. You can heat the oil or lard directly in the one saucepan, but having a water bath is safer.
4. Dissolve 1 oz sodium hydroxide in 7 oz of water in another container or cup. Be very careful. The reaction generates considerable heat. Add the sodium hydroxide slowly, bit by bit, and continuously stir to dissolve. Wear rubber gloves and safety glasses.
5. Slowly and carefully pour this solution into the melted lard. Mix thoroughly by stirring with a stick blender (or wooden spoon, which will take a lot longer).
6. You should notice that a solid starts to form. Continue to heat gently for another 10 minutes, stirring with the blender and occasionally stopping it and stirring manually to mix thoroughly. It should thicken.

A water bath allows gentle heating.

7. Option: this is where you can add food coloring or perfume to your soap.

   Add the coloring before you filter the soap flakes. Alternatively, experiment with plant pigments. For example, onion skin tea makes the soap yellow.

   Add perfume to your soap. This can be done in two ways. Petals of roses or other perfumed flowers can be layered over the flakes as they dry. Alternatively, add a few drops of perfume from a bottle into the mixture as the soap is being made.
8. Remove the small saucepan from the water bath and allow it to cool. If little or no soap has formed add 1 cup of salt solution to help precipitate the flakes. This latter step is seldom necessary.
9. The soap flakes or lumps can now be filtered through a kitchen tea strainer.
10. Wash the soap several times with iced water (this removes excess caustic soda).
11. Push the flakes into a dish or mold and allow to dry and set. It may take overnight to harden.
12. When hardened, remove the soap bar from the mold, and test the soap for its ability to lather in water. Cut the soap into smaller, useable blocks unless the molds are already the right size.

Homemade laundry detergent

Laundry soap or detergent is easy to make. Even though soap is different from a detergent, we don't differentiate here. You can choose to make a powder form (more like a soap) or a liquid form (more like a detergent), but liquids are preferred. If you use the powder form, add this to the water in the machine to dissolve before you add the clothes. If the powder is not fully dissolved you may get powder marks on your clothes. It is always best to make the solution.

This recipe requires the use of borax, which may cause controversy. Borax is sodium tetraborate (also sodium borate), a mineral found freely in deposits in the ground. It is alkaline, and should not be confused with boric acid, which has completely different properties. Borax can be poisonous in very high doses (so it should never be eaten in foods), but is unlikely to cause any adverse reactions to people who use it to wash clothes. It has the same health rating risk as washing soda, which is also used in this recipe.

Washing soda is sodium carbonate and should not be confused with baking soda (sodium hydrogen carbonate). However, if you cannot source washing soda then you can convert baking soda into washing soda by heating in an oven at 400°F for about one hour. Carbon dioxide and water are given off, so open the oven door every now and again, and stir the powder on the tray.

Ingredients

½ cup of borax

1 cup of washing soda

1 bar of pure soap (or equivalent amount of soap flakes)

2.5 gal water (and bucket)

Method

1. Shave or grate the bar of soap.
2. Add this to a saucepan with 0.5 gal of water. Heat and stir to dissolve soap flakes. If it does boil then lower the heat source and simmer until you have dissolved all the soap.
3. When cool, pour this solution into a 2.5 gal bucket, preferably one with a lid.
4. Add the borax and washing soda. Fill the bucket with water until near the top, and stir until everything has dissolved and mixed thoroughly.
5. Use one cup of this mix in every load.

Option: if you decide just to mix the dry powders together, then make sure the soap flakes are grated very finely and use about 1½–2 tbsp of powdered detergent for every load.

Powder detergent can be made rather than a solution.

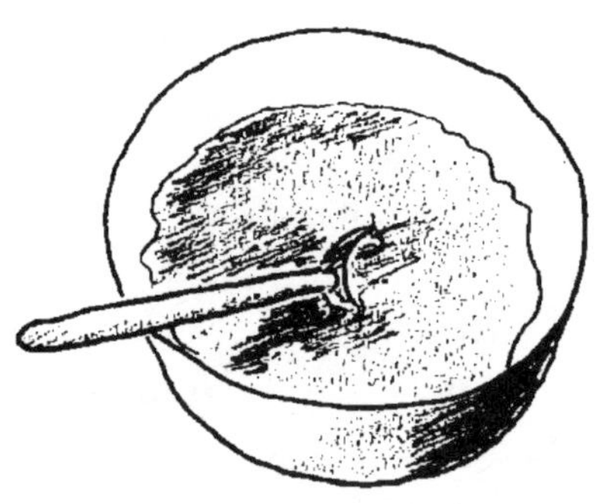

Stir thoroughly to dissolve the soap flakes.

Splitting bamboo

Bamboo is one of the strongest timbers, and the culms (poles or stalks) are used throughout the world in construction, building homes, reinforcing concrete, furniture, scaffolding and flooring. Bamboo is a multifunctional plant. So, besides the structural uses, some are used as food or fodder and others as shelter or a windbreak. Most bamboo species can be easily split as the wood "fibers" run longitudinally (and not radially), but some species are thick walled and require a lot more effort.

Bamboos are generally grouped into two types — clumping and running. Clumping varieties are the best for a backyard garden, and these tend to grow naturally in warmer tropical and subtropical areas.

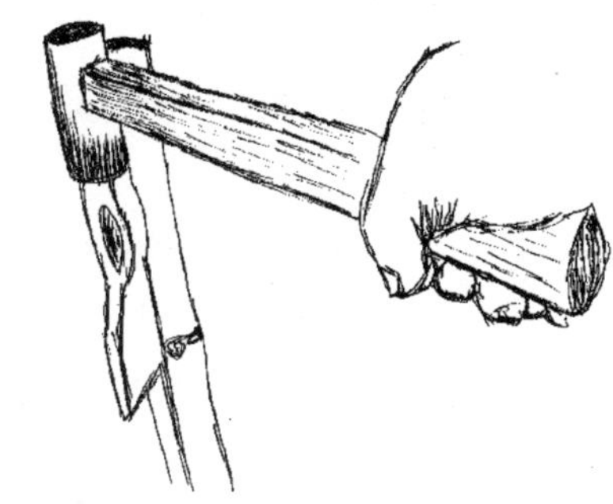

A wedge will need to be used for large bamboo.

Running bamboos generally come from temperate climates and are well suited to colder conditions. However, they can become rampant and spread throughout the garden (and into your neighbor's yard). Running bamboos are best kept in pots where they can be managed.

The culms to be split need to be somewhere between three to seven years of age, as these have the most lignin and strength. Fresh ("green") bamboo works best as dry bamboo sometimes tends to be a lot harder and more difficult to split. Besides, dried bamboo may already have splits in some places, and possibly not where you want. Always cut the culm and culm slats in half.

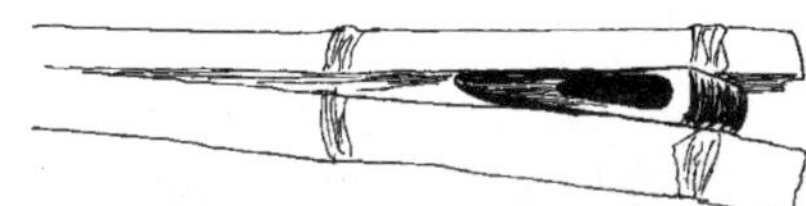

An ax head is a good wedge to split bamboo.

To split large bamboo culms (4 in or more in diameter), you need a machete or meat cleaver or a wedge (like an ax head).

The wedge is placed in the center of the culm and hit with a sledgehammer or gympie (bricklayer's hammer). As it splits you should place the cutting blade across the diameter (width) of the culm so that both sides split at the same time, essentially cutting the culm in half lengthwise.

This process can continue to further split the bamboo into slats and then slivers.

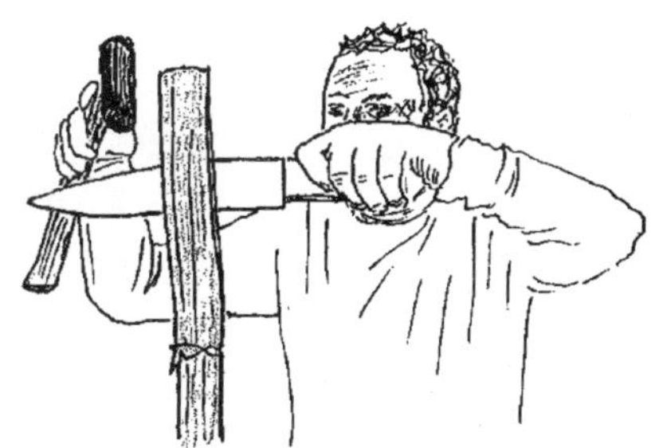

A smaller knife cuts thinner bamboo culms.

To split smaller bamboos you can use a thick knife and a hammer. Place the blade across the center of the culm and

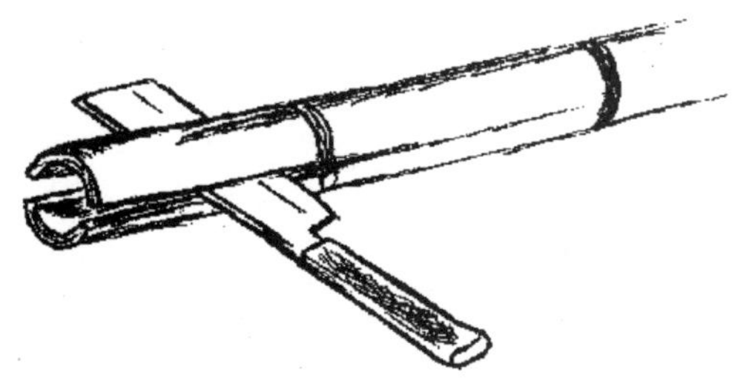

Thinner walled varieties can be easily split.

gently hit it with a hammer to start the split. Then use the knife to push downwards, moving the blade side to side to further cause splitting. You may need a little extra push or twist when you encounter a node.

For very thin sections use a utility knife. Start the slit and then push the bamboo through the blade (don't drag the knife, push the bamboo).

How to make playdough

Kids love playdough, and unfortunately they love to eat it too, so it is important to make a nontoxic product. This recipe does require heating but is easy to make. For gluten-free playdough use rice and corn flour instead of wheat flour.

While you can find recipes that require no cooking or cream of tartar, the playdough made without these is less pliable and cohesive, and more crumbly and dry.

Heating causes the proteins in the flour to interact to become a stretchy, elastic mass.

The cream of tartar is often used to stiffen the mixture and provide some stability to the product. The oil acts as a lubricant, making the playdough soft.

Ingredients

1 cup wheat flour (or ½ cup each of rice and corn flour)
¼ cup salt
2 tbsp cream of tartar (potassium hydrogen tartrate)
1 tbsp vegetable oil (mild odor, not like olive oil)
1 cup water
Food coloring — yellow, red, green, blue

Method

1. Mix dry ingredients in a small saucepan.
2. Place the oil, water and a few drops of one type of food coloring in a bowl. Blend.

3. Add the mixed liquid to the flour mixture in the saucepan. Mix thoroughly.
4. Heat the saucepan on a stove for a few minutes until the mixture thickens. Remove from heat.
5. When cool, turn out and knead on a lightly floured bench top. Kneading will improve texture.

   Add additional flour or water to make right consistency, although this is seldom necessary.
6. Repeat this procedure but use a different food color.
7. Store in a sealed plastic container or a sturdy plastic bag that can be zipped to prevent the playdough from drying out.

Playdough — make different-colored balls

Saving seeds

We have briefly discussed seed saving in earlier chapters, but it is worthwhile to expand on this here. Everyone should save the seeds from the strongest and most prolific vegetable plants they grow each year. Saving seed doesn't necessarily mean storing seeds for years, as the seeds from most plants have longevity of only a year or two.

Recalcitrant seeds, from subtropical and tropical plants, often have a viability of a few months, so those are typically planted as soon as they are picked.

You must remember that seeds are living things and eventually die if not planted. They have a finite lifetime. While many people tend to grow heirloom, open-pollinated plants, you can save the seed from hybrids.

Hybrid varieties are produced from crosses between different plants, and are bred for particular characteristics, such as good tasting fruit, disease resistance and bearing fruit earlier in the season.

Hybrid seeds are not the same as genetically modified organisms (GMOs), which tend to be produced so that they do not have viable seed, so new seed has to be purchased each year.

We are not going to discuss the seed saving techniques of individual plants, just make some general comments, so you can find out about the particular vegetables and fruits you grow when you need to.

» DID YOU KNOW?

Pollination is the process whereby pollen (male cells) is transferred and fertilizes the eggs of the female part of the flower. Seeds develop from fertilized eggs. If this process occurs on the same flower, then this is called self-pollination. If pollen is transferred from one flower to another, either on the same plant or between different plants, then cross-pollination has occurred. Pollination often needs insects, other animals (birds, mammals), water or wind to enable the pollen to move across to the female part.

Seed saving tips

- Some seeds can be harvested by securing a paper or plastic bag around the seed head and waiting until the pods open and seeds are ejected. This is a technique for the brassicas (broccoli, cabbage).
- Some seeds can be harvested by placing a large tarpaulin under the tree and then shaking the tree to dislodge seeds or their pods. This is a useful strategy for nuts.
- Seeds should be cleaned before storage. Remove any sticks, pods, leaf material. You may be able to use screens and sieves to accomplish this.
- Once harvested, seeds are often left to dry before storage. Larger seeds take longer to dry (a few days).
- Don't dry seeds directly in sunlight. They tend to desiccate too much and die.
- Store seeds in paper bags or glass jars. Don't place seeds in plastic bags as this causes the seeds to "sweat" and encourages mold to grow. Store in a cool, dry, dark place.
- There is no need to exclude air completely. Seeds, containing living cells, need a little bit of air to survive. While there is no need to put wax over jar lids, reducing air exposure prolongs seed life.
- It is important to exclude moisture from seeds as this causes fungus to grow, and to keep insects and other organisms out so that the seeds are not eaten. You can place a silica gel sachet in the jar to help control excess moisture.
- Seeds from some plants, such as tomatoes, are found in a mushy pulp. They need to be separated from this, and you can either wash the pulp and seeds in a bucket or scrape the seeds out as best you can. When washing, the pulp

Store seeds in paper bags, and label.

may float and the seeds will sink, where they can easily be retrieved. After scraping the seeds from the pulp, spread them over newspaper to dry out.

- Seed longevity is very sensitive to humidity (moisture in the air) and temperature changes. As a general rule, if humidity and temperature increase, then seed viability decreases.
- Some seeds can be stored in the freezer. You cannot store desiccation-intolerant (recalcitrant) seeds in a freezer as this destroys seed tissue when water expands as it freezes.
- Smaller seeds, which generally have a long viability, are ideal candidates for freezer storage, as long as they are dried correctly. Seeds stored in a freezer can be kept for longer periods.
- Consider swapping seeds with others. You can join a seed saving network or donate to a local seed bank. In this way, your seeds will be grown and good varieties dispersed throughout the neighborhood, and you can receive varieties that you might enjoy too.
- If you collect heirloom (traditional, heritage) seed varieties then they are mostly true to type. This means that they will produce plants that are the same as the parent plant.
- Hybrid seeds may not produce identical plants as their parent, so you may get a variety with slight differences in color, flavor or production.
- When seeds are stored make sure you label the container with information such as date of collection, place of collection (if not from your own garden), scientific name, common name and variety.

Community groups could set up a seed bank.

Seed saving is a rewarding task. Besides saving some money each year, there is great satisfaction in witnessing the cycle of nature.

Growing the seed, harvesting the plants for food or other uses and products, saving the seed and then regrowing them in the years that follow completes the natural cycles of nature.

Weaving (fiber basketry)

Very few people ever become rich as basket weavers. However, the materials you use are almost free (many fibers from your garden plants) and as long as you have time, basketry opens up another world of materials, plants and artistic expression.

Suitable plants include the leaves of cordyline, banana, dracaena, flax, gladiolus, iris, pampas grass, watsonia, dianella, gymea lily, lomandra, bulrush, sedge and carex. The stems and twigs of willow, wisteria, grapevines, couch grass, honeysuckle and jasmine are also commonly used.

Fresh leaf and twig material is best, as it tends to be soft and pliable enough to pass around poles and each other. You may need to soak some leaf material for some hours or even overnight to permit weaving to occur. Dampen off the leaves with a cloth before you start.

There are lots of books about weaving, many providing patterns and clear instructions on how to make baskets, hats and many more items. Only a brief outline of some basic techniques is presented here.

There are many patterns for weaving, but here are three common ones that can be used to make simple baskets, hats and even garden fences.

Square weave

This is a basic weave and is used as a base for bowls, hats, baskets and mats.

Method

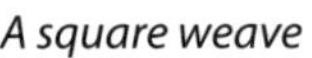

A square weave

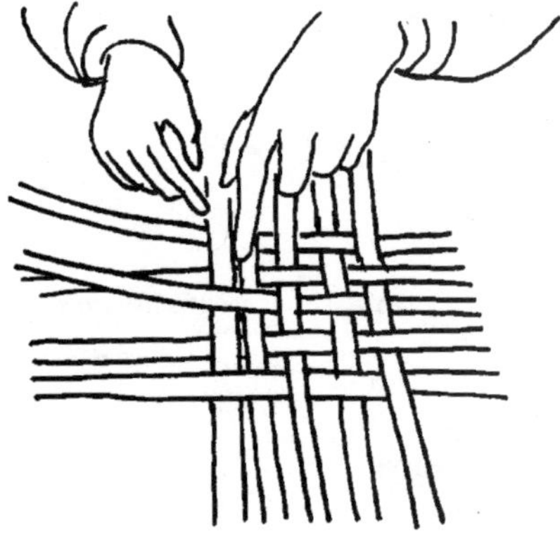

1. Lay leaves parallel to each other, with spacing the width of a typical leaf.
2. Place a heavy piece of wood or brick on one end to keep that end still.
3. Lift up every alternate leaf and place a new leaf across the remaining fibers.
4. Lift up the next alternating set and repeat the process.
5. Keep the overlapping leaves close together. It is possible to adjust the weave by pushing the fibers together to tighten it up.
6. To finish the "square" you could sew around the perimeter to hold the leaves together.

A variation of the square weave is the diagonal weave, which adds variety to your project.

Randing

This is a simple under-and-over weave, where fibers are wound through stakes. This weave is ideal for bamboo and brushwood fences where stem material is passed through poles.

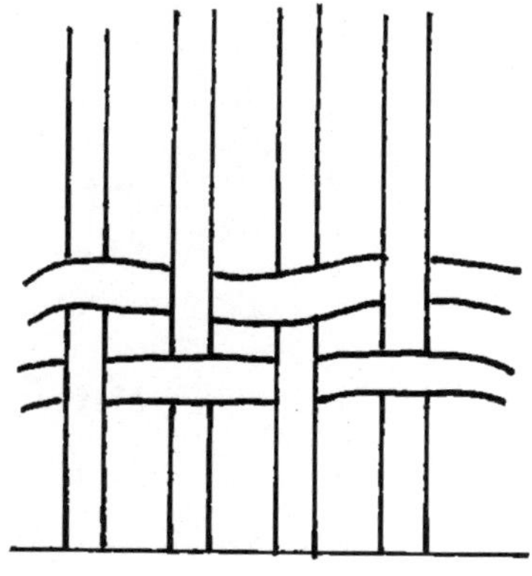

Randing

Method

1. Place stakes or poles evenly apart.
2. Weave the fiber in and out, as close as possible.
3. For the next layer, start the fiber in the opposite position.

Twining

Twining or pairing is a closer weave than randing. It is also an under-and-over weave but two fibers are threaded at the same time so that they cross each other.

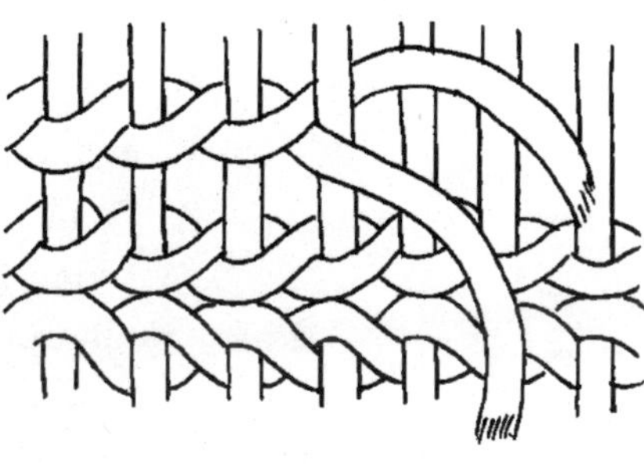

Twining

Method

1. Place stakes or pole evenly apart.
2. Take a pair of fibers. Thread one around and through the poles.
3. The second fiber starts on the opposite side of the stake and is threaded under and over the first fiber, so that they cross between the stakes.
4. Once threaded, push the pair of fibers down to sit firmly on the previous weave.

» DID YOU KNOW?

Rushes, reeds and sedges all make good material for weaving. All three are wetland plants and species of each type are commonly found throughout the world. Sedges are flat-bladed plants, rushes have cylindrical leaves (often with air spaces throughout), while reeds are from the grass family and include species of papyrus, carex, typha and phragmites.

Making rope from plant material

Normal rope is usually three strands of material wound around each other. This is not the same as a braid or plait, where the strands are woven together.

Using plant material to make braided rope is easy. It is similar to weaving. The simplest rope is a weave of three fibers, but thicker rope can be made by using five or seven strand plaits. The fibers of sisal, hemp and jute are commonly used to make rope but many other plants can be used. Long strips of flat-bladed plants such as banana, sedges, carex, watsonia and dianella are suitable.

Method

1. Obtain thin strips of leaves. Either cut the leaves with scissors or a utility knife into longitudinal strips (along the length of the leaf). You can also learn how to extract the fibers from the bark of some types of dead trees or use animal hair or sinew instead. If you want to practice the technique before you thrust into plant braiding, just use string.

2. Tie the three strands together at one end (with a simple knot, elastic band etc.).
3. To plait or braid, place two strands to one side. Always move the farthest strand (that is on the side with two strands) and pass it over the other (always across the center one).
4. Repeat this procedure until you have enough length in the rope. Tie the ends together so it doesn't unravel.
5. When you run out of plant material just overlap one strand about 2–3 in from the end and underneath of the current piece, enough to enable a few braids to keep the new material in place.

Three strand braid or plait

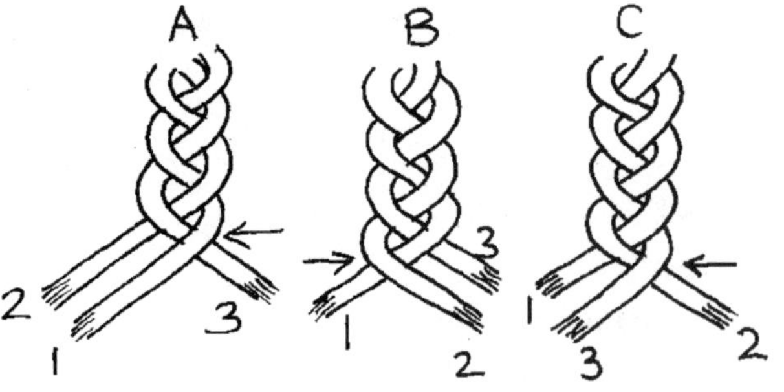

It is a bit tricky to hold the two overlapping pieces as one but you will get the hang of it.

You can use this braided rope to tie your bamboo fence or poles together, make a handle for your basket, produce a bracelet or leash, or for any other types of lashings.

Braiding allows you to join the leaves of onions or garlic together so that they can be hung as they dry out or stored until you need some for cooking.

Braid onions together and hang to store.

How to make candles

Most candles that you buy in shops are made from paraffin wax. Paraffin is produced from petroleum as a by-product from oil refining. Occasionally you can buy candles made from beeswax, but many other waxes are suitable, such as soy and palm and many combinations of any of these.

All candles need wax that has a low melting point so as the flame ignites the wick, the wax slowly melts (and burns) and exposes more wick.

Eventually, all that is left is a small pool of melted wax, which solidifies and can be reused to make more candles. Please read the whole method before you start, as there are options and hints to help you decide how to best proceed.

Materials required

Wax — soy, paraffin or beeswax

Twin saucepans — one smaller to fit inside the larger — for a water bath

Wick — braided cotton, paper core or zinc core, hemp cord

Mold — tin cans, ceramic jars, glass. You can purchase molds that you recover and use again or you can simply pour the wax into a glass or other container and leave it complete.

Option: dye or essential oil for scent

Method

1. Prepare the mold with the wick suspended centrally. The wick needs to be kept in place when the molten wax is poured into the mold, so some wicks you can buy are zinc core and these stand upright already. Cotton and paper braid wicks are reasonably rigid too but these and other types of wick cord can be supported by wrapping the end around a pencil or other object on top of the mold. The base of the wick needs fixing in some way. You can buy pre-made wick clips but you can fix the end by placing a small dob of silicone glue

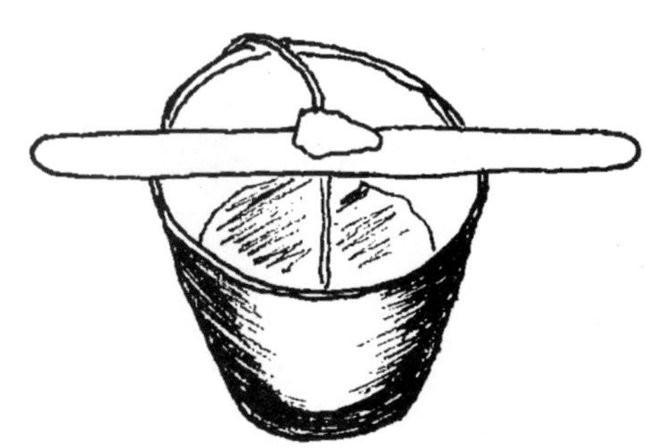

Suspend the wick and center it in the mold.

(or hot gun glue) attaching the wick to the mold and allowing this to set overnight (see also step 6).

2. Work out how much wax is needed to make your candle. If your candle is a cylinder, measure the mold diameter (then half this for the radius) and height (both in centimetres) and use the formula:
   $v = \Pi r^2 h$
   where v = volume, Π (pi) = 3.14, r = radius and h = height
   As an example, if the mold had a diameter of 2 in (radius = 1 in) and a height of 5 in then the volume of wax, v = 3.14 × 1 × 1 × 5 = 15.7 in^3*. As wax has a density of about .55 oz(in*^{3}*) you will need about 8.6 oz (1oz = 28g).*
3. Place your wax, cut up as best you can into smaller pieces, into the smaller saucepan, and place this inside the large saucepan half-filled with water. Heat the water to melt the wax. While you can carefully heat the wax in the single saucepan directly on the stove there is a very high risk that the wax will catch fire. The water bath will provide the usual 140–175°F heat that various waxes need to melt.
4. Option. When the wax has melted you can add a scented oil or dye, but they need to be oil-soluble and not only soluble in water. Only about 10% of essential oil, at most, should be added. You only need about 2 cups of dye. Add a little dye, mix and see what color it is. Add more until you see the tone of color you want. Mix thoroughly as the oil may separate from the wax (different densities).
5. Pour your wax into the mold. If you can preheat the mold (in low temperature oven) then the mold and wax will cool at the same temperature. If hot wax is poured into a cold mold then it starts to set straightaway, and possibly unevenly.
6. If you purchase stiff wicks then you can place the wick into the cooling wax in the mold. Just as the wax makes a soft skin on the top push the wick downwards to the bottom and endeavor to center it.
7. Because wax shrinks when it cools you may have to melt some additional wax and fill the mold to the top again.
8. When completely cooled (depends on the type of wax — anywhere from half-a-day to overnight) trim the wick to about ¼ in.

9. Removing the candle from the mold can be difficult. Here are some options to try:
   (a) You can purchase a mold release spray. Spray mold at step 1.
   (b) Place the candle into a refrigerator for 10 minutes. Gently tug on the wick and it should release.
   (c) Place the mold into a saucepan with hot water. The heat on the mold should be enough to slightly melt some of the wax, enough to enable you to pull the candle out.
10. Leave the candle to set for at least 2 days before you use. If you are giving your candle as a gift, label it with the ingredients you have used.

Candles made in glasses and jars

We could have discussed a lot of other useful skills that need to be reintroduced again, and these include mud brick making, sharpening tools, mixing and screeding concrete, making household cleaning and bathroom products (moisturizers, shampoos and toothpaste), grafting and budding, and making rammed earth walls and paths. Maybe in another book!

CONVERSION FORMULAE

1 cm = 0.394 in
1 in = 2.540 cm

1 m = 3.370 ft
1 ft = 0.305 m

1 km = 0.621 mi
1 mi = 1.609 km

1 ha = 2.471 ac
1 ac = 0.405 ha

1 g = 0.035 oz
1 oz = 28.350 g

1 kg = 2.205 lb
1 lb = 0.454 kg

1 tonne = 0.984 ton
1 ton = 1.016 tonne

1ml = 0.035 fl oz
1 fl oz = 28.413 ml

1 l = 1.760 pints
1 pint = 0.568 l

1 cup liquid = 240 ml

1 cup butter/sugar = 225 g

1 cup flour = 150 g

1 cup grated cheese = 110 g

1°C = 33.8°F
1°F = -17.2°C

ACKNOWLEDGMENTS

A SINCERE THANK YOU goes to the many people who have made valuable contributions to this work, and in particular:

Foreword: Rob Hopkins

Initial reading of chapters: Jenny Allen, Albert Bates, Graham Bell, Graham Brookman, Josh Byrne, John Champagne, Robin Clayfield, Naomi Coleman, Stewart Dallas, Chris Dixon, Darren Doherty, Ben Falloon, Julie Firth, Robyn Francis, Russ Grayson, David Holmgren, John Kitsteiner, Greg Knibbs, Kym and Georgie Kruse, Geoff Lawton, Ben Law, Ian Lillington, Max Lindegger, Penny Livingston, Cecilia Macauley, Hannah Moloney, Bill and Lisa Mollison, Rosemary Morrow, Jeff Nugent, Charles Otway, Scott Pittman, April Sampson-Kelly, Allan Savory, Craig Sponholtz, Sally Wise, Bruce Zell, Ken Yeomans.

Additional reading and comments: Adrian Baxt-Dent, Vicki Boxell, Laura Jones, Annora Longhurst, Ellen Krogdahl-Davie, Jo McLeay, Mignon Mitchell, Melissa Pizzato, Rachel Pontin, Martine Rousset, Kelly Thorburn.

Artwork by Simone Willis

Additional artwork: Chapter 9 — Kristin Scali

Proofreading: Josephine Smith — WordSmithWA

INDEX

Page numbers in *italics* indicate tables.

ABOUT THE AUTHOR

Ross Mars is a highly regarded permaculture teacher, designer and consultant. He is author of landmark *The Basics of Permaculture Design* and three other permaculture books, as well as two DVDs on energy efficient housing design and renewable energy systems. Ross manages Candlelight Farm, a permaculture demonstration site and training center in Western Australia. Over the past decade, he has delivered dozens of basic, design, advanced and diploma-level Permaculture Courses. Both a scientist with a PhD in Environmental Science and an entrepreneur, Mars also manufactures and supplies greywater and rainwater tank systems, and installs waterwise gardens and water-sensible irrigation systems.

A Note About the Publisher

New Society Publishers (www.newsociety.com), is an activist, employee-owned, solutions-oriented publisher focused on publishing books for a world of change. Our books offer tips, tools, and insights from leading experts in sustainable building, homesteading, climate change, environment, conscientious commerce, renewable energy, and more — positive solutions for troubled times.

The interior pages of our bound books are printed on Forest Stewardship Council®-registered acid-free paper that is 100% post-consumer recycled (100% old growth forest-free), processed chlorine-free, and printed with vegetable-based, low-VOC inks, with covers produced using FSC®-registered stock. New Society also works to reduce its carbon footprint, and purchases carbon offsets based on an annual audit to ensure a carbon neutral footprint. For further information, or to browse our full list of books and purchase securely, visit our website at: www.newsociety.com

New Society Publishers

ENVIRONMENTAL BENEFITS STATEMENT

For every 5,000 books printed, New Society saves the following resources:[1]

36	Trees
3,256	Pounds of Solid Waste
3,583	Gallons of Water
4,673	Kilowatt Hours of Electricity
5,919	Pounds of Greenhouse Gases
25	Pounds of HAPs, VOCs, and AOX Combined
9	Cubic Yards of Landfill Space

[1]Environmental benefits are calculated based on research done by the Environmental Defense Fund and other members of the Paper Task Force who study the environmental impacts of the paper industry.